Lecture Notes in Computer Science 5960

Commenced Publication in 1973
Founding and Former Series Editors:
Gerhard Goos, Juris Hartmanis, and Jan van Leeuwen

Xiaoyi Jiang   Matthew Y. Ma
Chang Wen Chen (Eds.)

Mobile
Multimedia Processing

Fundamentals, Methods, and Applications

 Springer

Volume Editors

Xiaoyi Jiang
University of Münster
Department of Mathematics and Computer Science
Einsteinstrasse 62, 48149, Münster, Germany
E-mail: xjiang@uni-muenster.de

Matthew Y. Ma
Scientific Works
6 Tiffany Court, Princeton Junction, NJ 08550, USA
E-mail: mattma@ieee.org

Chang Wen Chen
State University of New York at Buffalo
Department of Computer Science and Engineering
201 Bell Hall, Buffalo, NY 14260-2000, USA
E-mail: chencw@buffalo.edu

Library of Congress Control Number: 2010923704

CR Subject Classification (1998): C.2, H.5.1, H.3-5, D.2, I.2

LNCS Sublibrary: SL 5 – Computer Communication Networks
and Telecommunications

ISSN   0302-9743
ISBN-10   3-642-12348-1 Springer Berlin Heidelberg New York
ISBN-13   978-3-642-12348-1 Springer Berlin Heidelberg New York

springer.com

© Springer-Verlag Berlin Heidelberg 2010
Printed in Germany

Typesetting: Camera-ready by author, data conversion by Scientific Publishing Services, Chennai, India
Printed on acid-free paper   06/3180

Preface

The portable device and mobile phone market has witnessed rapid growth in the last few years with the emergence of several revolutionary products such as mobile TV, converging iPhone and digital cameras that combine music, phone and video functionalities into one device. The proliferation of this market has further benefited from the competition in software and applications for smart phones such as Google's Android operating system and Apple's iPhone App-Store, stimulating tens of thousands of mobile applications that are made available by individual and enterprise developers. Whereas the mobile device has become ubiquitous in people's daily life not only as a cellular phone but also as a media player, a mobile computing device, and a personal assistant, it is particularly important to address challenges timely in applying advanced pattern recognition, signal, information and multimedia processing techniques, and new emerging networking technologies to such mobile systems.

The primary objective of this book is to foster interdisciplinary discussions and research in mobile multimedia processing techniques, applications and systems, as well as to provide stimulus to researchers on pushing the frontier of emerging new technologies and applications.

One attempt on such discussions was the organization of the First International Workshop of Mobile Multimedia Processing (WMMP 2008), held in Tampa, Florida, USA, on December 7, 2008. About 30 papers were submitted from 10 countries across the USA, Asia and Europe. Fifteen papers were selected for presentation at the workshop covering various topics from image, speech and video applications, to multimedia services, and to navigation and visualization techniques. To put together this book, authors of selected papers from the workshop were invited to submit their extended versions. Several other scholars were also invited to submit their papers that cover additional important aspects of mobile multimedia processing that were not presented at the workshop. After a selection process involving two rounds of committee review followed by the editors' final review, we are delighted to present the following 15 chapters.

The first three chapters discuss how a mobile device can be used to assist accessibility and learning and what implementation issues there are in deploying various technologies in a mobile domain. In "Vibrotactile Rendering Applications and Mobile Phones," Shafiq Ur Réhman and Li Liu describe their work on sensing emotions through mobile vibrations and present a vibration coding scheme. In "Mobile Visual Aid Tools for Users with Visual Impairments," Xu Liu, David Doermann and Huiping Li develop several computer vision and image processing based applications on mobile devices to assist the visually impaired using a built-in camera. These applications include a color channel mapper for the color blind, a software-based magnifier for image magnification and enhancement, a pattern recognizer for recognizing currencies, and a document retriever for accessing printed materials. In "QR Code and Augmented Reality-Supported Mobile English Learning System," Tsung-Yu Liu, Tan-Hsu Tan and Yu-Ling Chu combine barcode and augmented reality technologies in a context-aware mobile learning application.

The next four chapters discuss issues related to extending client/server architecture and networking technologies to adapt to the mobile environment. In "Robust 1-D Barcode Recognition on Camera Phones and Mobile Product Information Display," Steffen Wachenfeld, Sebastian Terlunen and Xiaoyi Jiang present a robust 1-D barcode reader using a low-resolution camera phone to retrieve product information from the network. In "Evaluating the Adaptation of Multimedia Services Using a Constraints-Based Approach," José M. Oliveira and Eurico Manuel Carrapatoso develop a SMIL Media Adaptor for adapting multimedia services in mobile environments using a constraint algorithm based on a user's contextual changes. Further, in "Mobile Video Surveillance Systems: An Architectural Overview," Rita Cucchiara and Giovanni Gualdi present an overview of mobile video surveillance systems, focusing in particular on architectural aspects (sensors, functional units and sink modules). They also address some problems of video streaming and video tracking specifically designed and optimized for mobile video surveillance systems. In "Hybrid Layered Video Encoding for Mobile Internet-Based Computer Vision and Multimedia Applications," Suchendra Bhandarkar, Siddhartha Chattopadhyay and Shiva Sandeep Garlapati propose a Hybrid Layered Video encoding scheme that takes into account bandwidth characteristics and power assumption, and is shown to be effective for various mobile Internet-based multimedia applications.

The next three chapters present some representative mobile applications in entertainment. In "A Recommender Handoff Framework for a Mobile Device," Chuan Zhu, Matthew Ma, Chunguang Tan, Gui-ran Chang, Jingbo Zhu and Qiaozhe An present an electronic programming guide recommender system that enhances the user's TV experience in both the home and the new mobile TV environment. In "MuZeeker: Adapting a Music Search Engine for Mobile Phones," Jakob Eg Larsen, Søren Halling, Magnús Sigurðsson and Lars Kai Hansen adapt a music search engine for mobile phones and have conducted user evaluations for two prototype search applications: Web-based and mobile. In "A Server-Assisted Approach for Mobile Phone Games," Ivica Arsov, Marius Preda and Françoise Preteux propose a client-server architecture and a communication protocol for mobile games based on various parts of MPEG-4 that ensure similar user experience while using a standard player on the end-user terminal.

While we attempt to cover a wider range of mobile applications, we present some work on face and speech processing in the next three chapters. In "A Face Cartoon Producer for Low-Bit Digital Content Service," Yuehu Liu, Yuanqi Su, Yu Shao, Zhengwang Wu and Yang Yang focus on creating low-bit personalized face sketches and face cartoons from a photograph of a face without human interaction using a three-layer face modelling technique. In "Face Detection in Resource Constrained Wireless Systems," Grigorios Tsagkatakis and Andreas Savakis investigate how image compression, a key processing step in many resource-constrained environments, affects the classification performance of face detection systems. Further in "Speech Recognition on Mobile Devices," Zheng-Hua Tan and Børge Lindberg present an overview of automatic speech recognition (ASR) in the mobile context. They discuss three typical ASR architectures for mobile environment, their pros and cons, and compare the performance among various systems and implementation issues.

Finally, in the last part of the book we present two chapters dealing with the development environment of mobile device from the multimedia perspective. They aim at helping readers quickly build mobile prototypes from their research results. In "Developing Mobile Multimedia Applications on Symbian OS Devices," Steffen Wachenfeld, Markus Madeja and Xiaoyi Jiang describe how to develop a typical application on Symbian OS devices. Further, in "Developing Mobile Applications on the Android Platform," Guiran Chang, Chunguang Tan, Guanhua Li and Chuan Zhu explain how to deploy applications on the recently launched Android platform.

The intended readers of this book are primarily researchers wanting to extend traditional information processing technologies to the mobile domain or deploy any new mobile applications that could not be otherwise enabled traditionally. This book is dedicated to bringing together recent advancements of multimedia processing techniques for mobile applications. We believe that such a book will become a quality reference source of the state of the art in mobile multimedia processing and a valuable contribution to the research literature.

We thank the reviewers who have participated in our stringent reviewing process. They are:

Raj Acharya	Homer Chen
Antonio Krüger	Fatih Porikli
Sanghoon Sull	Patrick Wang
Kiyoharu Aizawa	Yung-Fu Chen
Yuehu Liu	Xiaojun Qi
Qi Tian	Hong Yan
Prabir Bhattacharya	David Doermann
Michael O'Mahony	Eckehard Steinbach
Steffen Wachenfeld	Tong Zhang

We would also like to extend our gratitude to all authors for contributing their chapters. Finally, our thank goes to Springer for giving us the opportunity of publishing this book in the LNCS series.

December 2009

Xiaoyi Jiang
Matthew Y. Ma
Chang Wen Chen

Table of Contents

Accessibility and Learning

Server Assisted Adaptation and Networking Applications

Entertainment

Face and Speech Processing

Development Environment

iFeeling: Vibrotactile Rendering of Human Emotions on Mobile Phones

Shafiq ur Réhman and Li Liu

Digital Media Lab.,
Umeå University, Sweden
{shafiq.urrehman, li.liu}@tfe.umu.se

Abstract. Today, the mobile phone technology is mature enough to enable us to effectively interact with mobile phones using our three major senses namely, *vision, hearing* and *touch*. Similar to the camera, which adds interest and utility to mobile experience, the vibration motor in a mobile phone could give us a new possibility to improve interactivity and usability of mobile phones. In this chapter, we show that by carefully controlling vibration patterns, more than 1-bit information can be rendered with a vibration motor. We demonstrate how to turn a mobile phone into a social interface for the blind so that they can sense emotional information of others. The technical details are given on how to extract emotional information, design vibrotactile coding schemes, render vibrotactile patterns, as well as how to carry out user tests to evaluate its usability. Experimental studies and users tests have shown that we do get and interpret more than one bit emotional information. This shows a potential to enrich mobile phones communication among the users through the touch channel.

Keywords: emotion estimation, vibrotactile rendering, lip tracking, mobile communication, tactile coding, mobile phone.

1 Introduction

Mobile phones are becoming part of everyday life and their prevalence makes them ideal not only for communication but also for entertainment and games. It is reported that mobile phone communication uptake in over 30 countries is exceeding 100% [1]. Current commercial mobile phones are becoming more sophisticated and highly intractable to all three major senses of human: *vision, hearing* and *touch*. Similar to the camera, which adds interest and utility to mobile experience, the touch system can improve interactivity and usability. Mobile vibration is one of the examples which clearly inform the user of incoming calls or SMS through a discernible stimulus, i.e. 1-bit information only. Nevertheless, it has been shown if the information is properly presented through mobile vibration, human can process touch information at the rate of $2 - 56$ bits/sec [2]; which makes it very interesting to explore the utility of vibration for mobile applications.

X. Jiang, M.Y. Ma, and C.W. Chen (Eds.): WMMP 2008, LNCS 5960, pp. 1–20, 2010.
© Springer-Verlag Berlin Heidelberg 2010

It is noted that mobile vibration engages people at different emotional levels and provides them with new ways of communicating and interacting. Different from other types of rendering, vibration could provide for *'unobtrusive device interaction through our 'private' medium touch'*, which can be highly desirable in certain situations [3]. How to render haptic feeling with mobile phones has been a hot research topic recently. On a PDA keyboard screen, researchers have shown that the haptic stimuli increase the performance of the user [4]. Similar results are reported in mobile phone interaction by exploiting both vibrotactile and gesture recognition [5], [6]. Brown et al. developed "tactons" for communicating information non-visually [7,5]. The "tactons" have been suggested to use for several applications, such as, "feeling who's talking". Using mobile phone's vibration, Kim et al. [8] has proposed the users' response to a mobile car racing game in which the user is informed about the state of a car and the road condition with a vibrotactile signal. In [9], we demonstrated how to render alive football game by vibration on a mobile phone. In [3], researchers have presented haptic interactions using their own piezoelectric device instead of the mobile phone's own vibration motor. They have also pointed out some difficulties e.g. power consumption, mechanical ground and a better understanding of user's interest. There are some commercial companies, e.g., Immersion AB [10], advancing the use of touch. They provide a sensory experience that is more entertaining and complete. In [11,12], the usability of mobile phones in a smart application is explored. The method selected for current work also uses vibration stimuli which are produced across the entire device (i.e. mobile phone).

The current work provides useful information by exploring the possibilities and strength of today's mobile phones as haptic communication interfaces. By considering a complete study of rendering emotions on mobile phones, we try to cover all technical blocks, from perception to user tests, which we hope that it is a showcase of using mobile phones to rendering dynamic multimedia. This chapter is organized in eight sections. After documenting the vibrotactile perception and interfaces design issues in section 3, we present guidelines for developing a vibrotactile rendering application in section 4 and section 5 provides design details for an interesting tactile information rendering application. After providing our vibrotactile coding scheme development details in section 6, the user experiences are documented in section 7. Finally, the future directions are documented.

2   *iFeeling* Rendering System

More than 65% of the information is carried out non-verbally during face-to-face communication [13] and emotion information has a significant importance in decision making and problem solving [14]. Missing emotional information from facial expressions makes it extremely difficult for the visually impaired to interact with others in social events. The visually impaired have to rely on hearing to perceive other's emotional information since their main information source is the voice. Unfortunately, it is known that hearing is a rather lossy channel to get emotional information. To enhance the visually impaired's social interactive ability, it is important to provide them reliable, "on-line" emotion information.

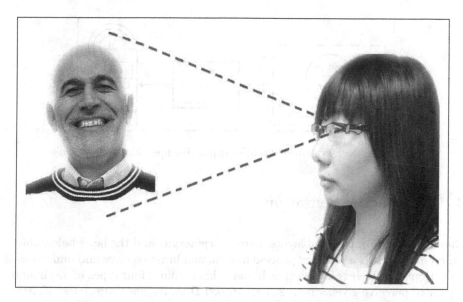

Fig. 1. An application scenario of our *ifeeling* rendering system

Today's mobile phones with current technology make them ideal for communication and entertainment. We can interact with a mobile phone with our three major human senses namely, *vision, touch* and *hearing.* Neither vision nor hearing is completely suitable for online emotional information rendering for the visually impaired: *visual rendering is useless for them; audio information is too annoying in daily life usage.*

Human touch provides a way to explore the possibilities of using vibration as emotional information carrier. In this chapter we demonstrate a system, *iFeeling,* where mobile vibration is used to render emotions. The application scenario of such a system is shown in Fig. 1; where the visually impaired wears a video camera pointing to the face of the person of interest, the captured facial video is analyzed and emotional parameters are extracted. Ongoing emotion is rendered by using the parameters on the mobile phone held by the visually impaired, who will interpret the emotion from vibration signals. The mobile phone produces vibrational signals corresponding to different facial expressions (expressive lips in current case). Obviously, mobile phones act as a social interface for the visually impaired to access online emotional information. The technical diagram is shown in Fig.2. The challenge is *how to display emotions through vibration of mobile phones.*

In order to make an effective vibrotactile system, one has to consider the human touch perceptual capabilities and limitations along with vibrotactile display design suitable for current mobile phones. In following sections, we provide a brief overview of touch perceptual capabilities along with foundation knowledge of vibrotactile actuator.

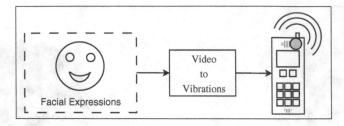

Fig. 2. A block system diagram of expressive lips video to vibration

3   Vibrotactile Sensation

Human skin is composed of layers, namely, *epidermis* and the layer below this is called *dermis*. It is a complex process how human brain receives and understands vibration stimuli. It is known that human skin employs four types of mechanoreceptor to perceive given stimulus, i.e., *Merkel Disk, Ruffini Corpuscles, Meissner Corpuscles* and *Pacinian Corpuscles* (see Fig. 3) [16]. Afferent fibber branches that form a cluster of Merkel's discs are located at the base of a thickened region of epidermis. Each nerve terminal branch ends in a disc enclosed by a specialized accessory cell called a *Merkel cell* (NP II). Movement of the epidermis relative to the dermis will exert a shearing force on the Merkel cell. The Merkel cell plays a role in sensing both touch and pressure. Ruffini Corpuscles (NP II) are spray-like dendritic endings (0.5-2 mm in length) in the dermis hairy skin and are involved in the sensation of steady and continuous pressure applied to the skin. Meissner Corpuscles (NPI) are found just underneath the epidermis in glabrous skin only, e.g., fingertips. Pacinian Corpuscles (PC's) are pressure receptors, located in the skin and in various internal organs. Each is connected to a sensory neuron. It can achieve a size of 1-4 mm. It has its private line to the CNS. Vibrotactile sensation generally means the stimulation of skin surface at certain frequency using vibrating contactors. Pacinian Corpuscles are considered performing a major role in vibrotactile perception with maximum sensitivity at higher frequencies, i.e., $200 - 300$ Hz.

The phenomenon of vibrotactile sensation was firstly introduced by Geldard [17]. There are four different parameters through which vibrotactile information coding can be studied, namely, *frequency, amplitude, timing* and *location*. A vibrotactile stimulus is only detected when the amplitude exceeds a certain threshold, known as detection threshold (DT). In general, DT depends on several different parameters [18,19,20] but mainly on the frequency ($\sim$ 20-500 Hz) and location (fingers are most sensitive 250-300 Hz). Vibrating stimulus up to 200 Hz results in synchronized firing of afferent nerve fibbers and considered capable of great physiological information flow, i.e., due to presence of large number of mechanoreceptor esp. PC's [21].

By using single detectable stimulus simple messages can be encoded [9]. J. van Erp has given detailed information for the use of vibrotactile displays

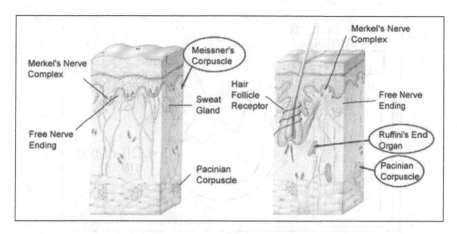

Fig. 3. Human skin cross section labelled with mechano-receptors [15]

in human computer interaction [22]. While presenting tactile coding information a designer should consider the following principals [23,24,25,26]:

- No more than 4 intensity (magnitude) levels should be used.
- No more than 9 frequency levels should be used for coding information and the difference between adjacent levels should be at least 20%.
- The time interval between adjacent signals should be more than 10 milli-seconds to prevent the "saltation effect".
- The vibration on the hand should be carefully dealt with and long durations might make users irritated.
- Vibration intensity can be used to control threshold detection and sensory irritation problem; similar to volume control in audio stimuli.

4 Vibrotactile Rendering

Normally, there are two commonly used display techniques to generate vibration: 1) *using a moving coil, usually driven by a sine wave*; 2) *a DC motor*, which is used in most mobile phones. Since we are using a mobile phone to deliver emotional information, the DC motor based vibrotactile display is used in the *iFeeling* system.

4.1 Vibrotactile Actuators

In almost all commercially existing mobile phones, a vibration actuator has already built in. It is natural to render emotion on a mobile phone with a vibration motor. The vibration patterns are generated by rotating a wheel with a mounted eccentric weight as shown in Fig. 4. The wheel rotates at a constant speed. Switching it on and off produces the vibration. The advantages by using such a vibrotactile actuator are

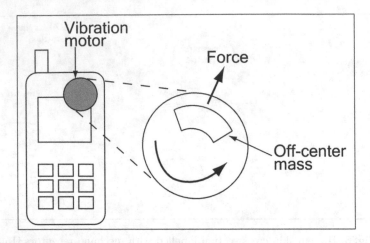

Fig. 4. A mobile phone with a vibrator [9]

- having light weight and small miniature size (dia ~ 8.0 mm),
- using low operating voltage(~ 2.7 − 3.3V),
- producing distinguishable tactile patterns,
- existing in most mobile phones and producing enough force,
- not power hungry contrary to other actuators [27,28,29]

Furthermore, it does not require any additional hardware to do rendering, which can result in increase in size or weight.

4.2   Vibrotactile Coding

To understand the vibration stimuli produced by a mobile phone vibrator, one must consider the input voltage and response vibration produced by the vibrator. Yanagida et al. [30] noted 100-200 msec response time of a vibration motor (see Fig. 5).

By properly controlling the motor response, one can create perceptively distinguishable vibrotactile patterns. Such an efficient technique is the **Pulse Width Modulation** (PWM) [9,30]. Pulse-width modulation uses a square wave whose pulse width is modulated resulting in the variation of the average value of the waveform. PWM allows us to vary the amount of time voltage is sent to a device instead of the voltage level sent to the device. The high latency of the vibration motor(i.e. the slow reactions to changes in the input signal)results in a signal which resembles those created by amplitude modulation. By switching voltage to the load with the appropriate duty cycle, the output will approximate a voltage at the desired level. In addition, it is found that PWM also reduced power consumption significantly compared to the analog output approach.

Thus, the vibration of an actuator can be determined by the on-off rotational patterns through PWM. The vibration can be controlled by two types of vibration parameters: 1) *frequency*; 2) *magnitude*. The vibration frequency is

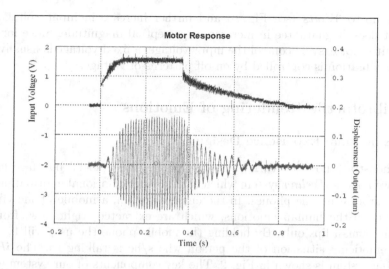

Fig. 5. A vibration motor response to stepwise input voltage reported by [30]

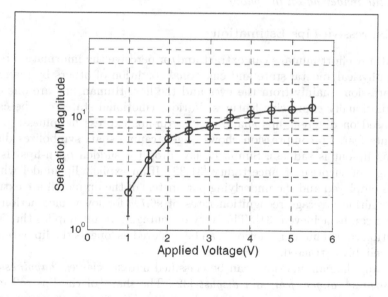

Fig. 6. Perceptive magnitude against input voltage reported by [29]

determined by the on-off patterns while the magnitude is determined by relative duration ratio of on-off signals. A vibrotactile stimulus is only detected when the amplitude exceeds a certain threshold, i.e., DT. To make sure that the generated vibrotactile signals are perceptually discernible, one has to carefully follow the principles of human's vibrotactile perception (see details in sec 6). Jung et al. [29] has shown a nonlinear mapping between the perceptual magnitude and input voltage. One can see that the perceptual magnitude increase almost linearly up to 3*Volts* (see Fig. 6) and further increase in input voltage over 3.5 Volt doesn't guarantee in increase in perceptual magnitude. Since for most phones it is not easy to control the input voltage, a fixed voltage is usually used and the vibration is controlled by on-off patterns of voltage.

5   Vibrotactile Rendering of Emotions

5.1   Emotional Experience Design

To demonstrate how to "turn" a mobile phone into a social interface for the blind, we build a *iFeeling* system which could render emotional information with the vibrator on mobile phones. In the current system, a mobile phone vibrates according to the human emotions, which are extracted, right now, from the dynamic human lips only. By holding the mobile phone the user will be aware of the emotional situation of the person who s/he is talking to. The *iFeeling* rendering system is shown in Fig. 2. The key components of our system are: *1) emotions from expressive lips 2) vibrotactile coding of emotional information 3) vibrotactile rendering on the phone.*

5.2   Expressive Lips Estimation

Without any effort humans can extract and/or perceive the information regarding the physical/mental state and emotional condition of others by reading facial expressions mainly from the eyes and the lips. Human lips are one of the most emotionally expressive features. Various emotional states can be categorized based on different lips shapes, e.g., grin-lips express happiness; grimace represents fear, lip-compression shows anger; canine snarl symbolizes disgust, where as lip-out is sadness. Similarly, jaw drop (or sudden open-lips) is a reliable sign of surprise or uncertainty [31,32]. If an explicit lip model with the states is employed and the underlying parameters of the lip model are extracted for recognition, average recognition rates of 96.7% for lower face action units from lips can be achieved [31]. This very encouraging result implies that a high recognition rate of human emotions can be achieved as long as the lip parameters can be reliably extracted.

Normally, human emotions can be classified into six classes: *happiness, sadness, surprise, anger, fear* and *disgust* [32]. The shape of the lips reflects the associated emotions. By examining the dynamics of lip contours, one can reliably estimate the following four types of emotions: *happiness, surprise, disgust* and *sadness*, simply due to that they are closely related to quite dissimilar shapes of the lips. Therefore, to estimate emotions from lips the key is lip tracking.

To enable an accurate and robust lip tracking, an intuitive solution is to find *'good'* visual features [33] around the lips for localization and tracking. The motion and deformation from tracked features are then used to deform the modelled semantic lip contours. Here we adopt the strategy of an *indirect lip tracking* [34]. Since it is not the semantic lips that will be tracked but the features around the lips, we call the approach *lipless tracking* [34]. This approach has the following unique features:

- Lip features are localized instead of semantic lip-model or lip-states [31].
- It is not dependent on large databases and heavy training [35].
- No initial manual operation and fully automatic tracking on a per user basis [36].

In *lipless tracking* [34], a 1D deformable template is used to represent the dynamic contour segments based on lip features. Personal templates based on four images per subject (normal, happy, sad and surprise) are taken off-line and segment contours based on features are identified manually. These lipless templates serve as look-ups during the lipless tracking process. In the scheme, Lipless Template Contours (LTC) are fitted to the edges extracted from individual video frames. The deformation flexibility is achieved by dividing the each template into smaller segments, which are split apart. This allows the templates to be deformed into shapes more similar to the features in the given video frames. Formally, the LTCs are divided into n sub-sequences or segments. For each sub-sequence, $i = 1, .., n$; t_i denotes the index to the first *'site'* in the ith segment. In other words, the ith segment contains the sites $m_{t_i}, ..., m_{t_{i+1}-1}$. The length of individual segments is fixed to approximately 5 pixel; i.e. the indexes $t_1, ..t_n$ where $t_1 = 1$, and for $i = 2... n$, t_i is defined by

$$t_i \equiv \min\{t = t_{i-1}, .., k : |m_t - m_{t_{i-1}}| \geq 5\} \tag{1}$$

where $|m_t - m_{t_{i-1}}|$ denotes Euclidean distance between the two sites.

When allowing one pixel gap or overlap, the overall deformation is around $1/5 = 20\%$ of the original template size. After template matching, the best matched template is selected. The search over the edges using lipless tracking finds the optimal location of the segmented templates. Expressive lips are estimated based on lip states (i.e. normal lip-state, happy lip-state (smile + laugh), sad lip-state and surprise lip-state).

5.3   Classification of Dynamic Emotions Information

After performing the template matching, two parameters are extracted to estimate certain emotions, i.e. *Emotion Type Estimator (ETE)* and *Emotion Intensity Estimator (EIE)*. ETE estimates the type of emotion based on matched LTC and EIE is based on how much LTCs are deformed from their original shape within allowed deformation (i.e. 20%). Formally, if u is given LTC and u^j is detected ETE at time t_i the EIE E^j is calculated as

$$E^j_{intensity} = u^j_{t_{i+1}} - u^j_{t_i} \tag{2}$$

Here $j = 1, 2, 3, 4$ corresponding to different LTC's.

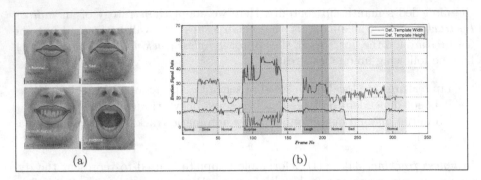

Fig. 7. Expressive lips : (a) Lip tracking based on 4-lip contours states from real-time video sequence. (b) Emotion indicators from template deformation. From template deformation parameters emotion type and emotion intensity are estimated.

From our live video collections, the results of expressive lips localization using lipless tracking are shown in Fig. 7(a). Fig. 7(b) shows emotion indicators from LTC deformation parameters of live video sequence, which will be used as the EIE. The details of method and techniques along with results both on our private collections and a publicly available database can be found in [34].

6   *iFeeling* Vibrotactile Coding Scheme

From *lipless tracking* two parameters namely *Emotion Type Estimator (ETE)* and *Emotion Intensity Estimator (EIE)* from expressive lips are extracted. The information is used to render emotions on mobile phones. In this section, we will provide details on how these two parameters are coded into vibrotactile information.

Primarily, there are four different ways to render vibrotactile information, namely, *frequency, amplitude, timing* and *location*. In our system, a vibrotactile signal is controlled by two parameters of vibration: 1) *frequency*; 2) *magnitude*. The *emotion type* is coded into the frequency of vibrotactile stimuli, whereas *emotion intensity* is coded into the magnitude of the vibrotactile pattern. For current application, we consider coding scheme which informs users about *"What is the type of emotion and what is its intensity?"*

In the design of vibrotactile coding of expressive lips, we developed self explaining tactile messages . As it is reported that 5-6 vibration modes are suitable for vibrotactile display in mobile phones because of human's limited touch recognition and memorizing capabilities [9]. In vibrotactile coding, four types of emotions (i.e. normal, happy (smile + laugh), sad and surprise) are estimated from lipless tracking and coded in vibrotactile signals.

In the vibrational signal, the information of the "emotion type" and its "intensity" is provided. To reduce the mental workload of the mobile users no vibration signal is given when the expressive lips are in the neutral state. This is the reason

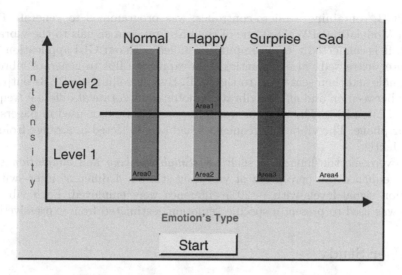

Fig. 8. Divided GUI areas for vibrotactile coding; Users are asked to indicate the emotion and its "intensity" (intensity level bottom-up)

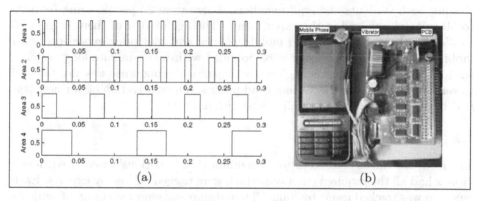

Fig. 9. (a) Vibrotactile modes used for expressive lips rendering on mobile phone. (b) A mockup of a mobile phone with the vibration motor driven by our self-designed PCB.

that the GUI was not divided into more areas (see Fig. 8). Fig. 9(a) shows the vibration stimuli for specific expressive lips to associate GUI areas.

6.1 Experimental Platform

To carry out a user test, a vibrotactile mobile system was built (although vibration can be also rendered on commodity mobile phones under the mobile Java platform, J2ME, for example, we would like to focus on more technical issues rather then practical implementation). Currently our system's tactile module is made up of two parts: *a vibrotactile box (VTB)* and *an ordinary mobile phone*. The VTB consisted of single printed circuit board (PCB) as shown in Fig. 9(b).

The PCB, containing a micro-controller, was programmed to generate Pulse Width Modulation (PWM) at one of the pins and send signals to the vibration motor. It is connected with a computer installed with our GUI application. The computer extracted video semantics from expressive lips to generate vibrotactile signals and then sent them to the PCB. By controlling the PCBs output to switch between *on* and *off*, the vibration motor could rotate at different frequencies and intensities. The vibrator was a low-cost coin motor used in pagers and cellular phone. The vibration frequencies had been selected in a range from 100 Hz to 1 kHz.

To overcome the limitation such as, *stimuli masking* and *adaptation* effect and to enhance the perception of vibrating stimuli, 4 different inter-switched vibration signal levels with ∼ 20% difference were employed. Each vibration signal was used to present a specific "emotion" estimated from expressive lips.

7   User Study

7.1   Participants

Participants were recruited from the campus of Umeå University, Sweden, including both students and staff members. There were 20 people aging from 20 to 50 in total involved in the experiments. The age of 70% of them is from 26 to 35. All the participants had mobile phones and had prior experience with mobile vibration. All the users were provided with a two-minute long training session. During the experiments, the participants hearing sense was also blocked to remove any auditory cues using head phones. The motivation of our user test had been clearly explained to all the participants.

7.2   Stimuli

The recorded videos of facial expression mimic (i.e. "emotions") were used. These videos had all the required emotions which were tagged off line by experts. Each emotion was tracked frame by frame. Time stamp and emotion stamp of subjects faces were recorded and used for the stimuli.

7.3   Procedure

First, we introduced the purpose of our experiments to the participants and explained how to use the GUI system. Each participant held the mobile phone with the vibration motor in one hand and a computer mouse in the other hand. To obtain sufficient experimental data, a vibrotactile sequence was given to the participant twice and the subjects were asked to point out the "emotion" based on vibration using mouse click (see GUI Fig. 8). The first round was carried out before the additional training process and the second round started after the training session. When the participants finished the first round and started to perform the second one, they were considered to be more experienced (because they had been trained by the first experiment). However, due to different stimuli, this did not affect the study of the effect of training process in the experiment. In the end, each participant was given a questionnaire for measuring levels of satisfaction and main effects. Each question was elaborated and discussed with the participants if required. After one-minute training session all the subjects took part two experiment sessions. The session had 10 minutes time duration approximately with 5-minutes time between each session for resting.

7.4 Effectiveness

Effectiveness is about whether or not a task could be correctly accomplished with the specified system. A simple indicator of effectiveness is the success to failure ratio for task completion [37]. In the current study, the error rate was calculated by comparison between the users' inputs and original data, i.e., the emotion and the timing stamped video sequences. When a certain emotion was displayed, e.g., in *surprise*, a vibrotactile signal was sent to drive the vibration motor. The user who experienced vibration was asked to use a computer mouse to click the area accordingly. The GUI application recorded the time and area when a mouse button was clicked.

The results of these experiments shows that performance for almost all subjects is increased with the passage of time. There are less false identification of areas (which represents emotions in GUI) based on the presented vibrotactile patterns (see Fig. 10, 11). It can be concluded that user perception of the system effectiveness has increased with training and so did its usability (see Tab. 1). The results also show that average false identification for designed coding scheme are decreased, i.e., 28.25% to 15.08%. The reason why less error in second session might be due to the more training. In Fig. 10,11, the delay time between original signal and user detection can be seen; it can be stimuli detection time plus user reaction time (as observed during experiments).

Table 1. Measured effectiveness of vibration

Mean	Before training	After training
Overall error rate	0.1875	0.0687
GUI Area1 error rate	0.1927	0.0511
GUI Area2 error rate	0.2025	0.1011
GUI Area3 error rate	0.1825	0.1014
GUI Area4 error rate	0.1723	0.0213

7.5 Efficiency

Efficiency is about how much effort is required in order to accomplish a task. An efficient system ought to require as little effort as possible [37]. In the experiments, we used the reaction time of a user as an indicator of efficiency. The

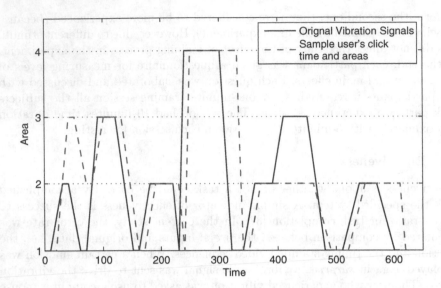

Fig. 10. Sample user results: before training

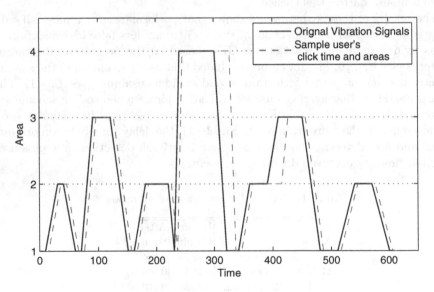

Fig. 11. Sample user results: after training

reaction time was measured by computing the delay between the time when a vibration signal was triggered and the time when the user clicked an area on the screen. The delay time contained three parts: *rendering time*, *cognitive time* and *move-and-click time*. Since both the rendering time and the move-and-click time were approximately constant, we assumed that the delay time was a reliable indicator of cognitive time.

Table 2. Measured efficiency in session 1

Mean (s)	Before training	After training
overall delay	0.6230	0.5202
GUI Area1 to GUI Area2 delay	0.4068	0.6173
GUI Area3 overall delay	0.6823	0.5134

Table 3. Measured efficiency in session 2

Mean (s)	Before training	After training
overall delay	0.8404	0.8069
GUI Area1 overall delay	0.6838	0.7534
GUI Area2 overall delay	0.7232	0.7693
GUI Area2 to Area1 delay	0.4060	0.2308

Measured efficiency (delay time) results for session 1 and session 2 are shown in Table 2 and Table 3. From the tables one can see that for both cases, the overall delay decreased by 8.2% and 4.0%, while Area1 and Area2 overall delay increased. After a careful examination of the delay results, we found that there was no stable pattern in delay time when a user made mistakes. To get a real picture of the duration of time a user needed to react to the stimuli, we noticed the delay time when the user made a right decision. The delay time with correct results in both sessions decreased. It implies that after training, users tended to take less time to make correct decisions but the training might have also confused the users as they users tended to use more time when making mistakes.

7.6 User Satisfaction

Satisfaction refers to the comfort and acceptability of the system for its users and other people affected by its use [37]. A useful measure of user satisfaction can be made if the evaluation team's measure is based on observations of user attitudes towards the system. Thus, it is possible to measure user attitudes using a questionnaire, e.g., *"Is this application interesting? 1=very boring (negative) to 7=very interesting (positive)."* The satisfaction levels of our experiments were measured by giving participants questionnaires.

At the end of the experiments, each participant was given a questionnaire to measure levels of satisfaction and main effects. Our subjective questionnaire used Likert style 7-point rating system, which scales from 1 to 7 [38]. Its values represented strong disagreement (negative) and strong agreement (positive), respectively. Fig. 12 shows mean questionnaire responses to questionnaire questions. Each label on the x axis in the figure represents a question where we used the following notation:

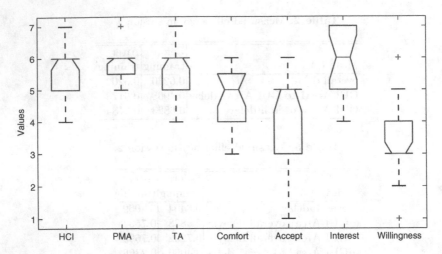

Fig. 12. Mean Questionnaire Scores

- HCI - *Is the human computer interface easy to use?*
- Pattern Mapping Accuracy (PMA) - *Is the pattern mapping recognizable?*
- Trainability (TA) - *Is the training helpful?*
- Comfort - *Is this application comfortable to use?*
- Accept - *Is this application acceptable?*
- Interest - *Is this application interesting?*
- Willingness - *Are you willing to buy such a service at a cost of 10 SEK per month?*

Experimental results indicated a high interest level for our application, mean score of 6.2311. Participants gave an average score of 5.95 to difficulty-easiness (to recognize the vibration signals) before training and an average score of 5.5532 after training, which considered the training a helpful procedure. An average rating of 4.7826 was given by participants to consider it a good application and showed a rate of 3.5621 on willingness to buy the service.

8   Concluding Remarks and Future Direction

The *iFeel* mobile system is part of our investigation to use mobile phones as an intuitive expressive interface for emotion rendering. It can be concluded from the application (discussed above) that mobile phones can be a creative and expressive devices as well as communication platforms. It is a solution that establishes and channels social interaction effectively just by a slight training cost. The user tests show that the subjects are able to recognize major emotions presented to them after very little training. The designed vibrotactile patterns are rather effective and usable but still there are some issues (needed further discussion), how to present more complex information and what will be the training cost for such a design. The results of this study suggest that a proper use of built-in vibrotactile actuator to produce tactile feedback can improve the usability and the experience.

Current commercial cell phone companies are considering innovative solution for normal users as well visually impaired. Samsung is working on a "Touch Messenger" mobile phone for the visually impaired. It enables visually impaired users to send and receive Braille text messages. The 3-4 buttons on the cell phone is used as two Braille keypads and text messages can be checked through the Braille display screen in the lower part. Similarly, HP and Humanware have launched a handheld PC (Maestro), the first mainstream handheld PC for blind and visually impaired persons. It is built on the HP iPAQ Pocket PC h4150 platform and has features like text-to-speech technology plus a tactile keyboard membrane over its touch screen for the visually impaired persons. It is based on Bluetooth wireless technology, and can be operated with or without an external keyboard (Braille or standard) [39]. Recently, Google engineers have introduced an experimental "Eyes-Free" touch interface for Android powered mobile phones which is based on an audio to text software library. The main feature is GPS use for navigation help [40]. Another research project is being developed by Cognitive Aid System for Blind People (CASBLiP) [41]. The CASBLiP project funded from the European Union(EU). There are a number of universities and blind institutions involved in the consortium. The proposed CASBLiP system is based on lasers and digital video images to create a three-dimensional acoustic map which, when relayed through headphones, enables users to "see" the world with sound. The signals received via headphones guide and assist the user to navigate around obstacles of the outside world. The user has to wear glasses with miniature video cameras mounted on them which provide the necessary video vision [42].

It can also be seen that the current mobile phone technology is capable of vibrotactile information communication without any major hardware modifications. We believe that the platform will promote not only existing normal mobile usage but also be beneficial for visually disabled persons. Our results show that today's mobile phones can be used to render more than one bit information. The current experimental studies provide straightforward and simple guidelines for designing exiting mobile phone applications by engaging user touch perception. Our research also highlights that a vibrotactile feedback is acceptable for social interfaces in the designed form, which means creative cross-modal applications can be developed based on the presented vibrotactile coding guidelines. In future studies, we will examine the issues regarding hardware design and more compact software solution for other mobile phone applications such as navigation help purely using tactile channel. While developing such a system, there are following critical issues which should be dealt with carefully i.e.,

 − energy consumption
 − actuator position
 − actuator material fatigue
 − vibration parameters
 − human tactile perception

9 Further Readings

We highly recommend the work done in the following research areas to the readers who want to learn more about the topic.

The Physiology of Touch: [43], [21], [44], [15] provide the fundamental knowledge of human touch sense and discuss its functionality in details.

Haptics: To read more about recent direction about haptics, readers are encouraged to read [45], [46], [47].

Vibrotactile Rendering in Mobile phones: [9], [48], [7,5] and [10] are recommended for further reading for vibrotactile rendering on mobiles.

References

1. Wallace, B.: 30 Countries Passed 100% Mobile Phone Penetration in Q1 (2006)
2. Kaczmarek, K., Webster, J., Bach-Y-Rita, P., Tompkins, W.: Electrotactile and Vibrotactile Displays for Sensory Substitution Systems. IEEE Transactions on Biomedical Engineering, 1–16 (1991)
3. Luk, J., Pasquero, J., Little, S., MacLean, K., Levesque, V., Hayward, V.: A role for Haptics in Mobile Interaction: Initial Design using a Handheld Tactile Display Prototype. In: Proceedings of CHI 2006, Interaction Techniques: Haptic & Gestural, pp. 171–180 (2006)
4. Brewster, S., Chohan, F., Brown, L.: Tactile Feedback for Mobile Interactions. In: Proceedings of ACM CHI 2007, San Jose, CA, pp. 159–162 (2007)
5. Brown, L., Williamson, J.: Shake2Talk: Multimodal Messaging for Interpersonal Communication. In: Oakley, I., Brewster, S. (eds.) HAID 2007. LNCS, vol. 4813, pp. 44–55. Springer, Heidelberg (2007)
6. Williamson, J., Murray-Smith, R., Hughes, S.: Shoogle: Multimodal Excitatory Interfaces on Mobile Devices. In: Proceedings of the ACM CHI 2007 (2007)
7. Brown, L., Brewster, S., Purchase, H.: Multidimensional Tactons for Non-Visual Information presentation in Mobile Devices. In: 8th conference on Human-Computer Interaction with Mobile Devices and Services, Helsinki, Finland
8. Kim, S.Y., Kim, K.Y.: Interactive Racing Game with Graphic and Haptic Feedback. In: Haptic and Audio Interaction Design
9. Réhman, S., Sun, J., Liu, L., Li, H.: Turn your Mobile into the Football: Rendering Live Football Game by Vibration. IEEE Transactions on Multimedia 10, 1022–1033 (2008)
10. Immersion: Immersion Cooperation, http://www.immersion.com/
11. Schmidt, A., Laerhoven, K.: How to Build Smart Appliance. IEEE Personal Communication, 66–71 (2001)
12. Siewiorek, D., Smailagic, A., Furukawa, J., Krause, A., Moraveji, N., Reiger, K., Shaffer, J., Wong, F.L.: SenSay: a Context-Aware Mobile Phone. In: Proceedings of 7th IEEE Int. Symposium on Wearable Computers (2003)
13. Argyle, M.: Bodily Communication. Methuen and Co., New York (1988)
14. Lisetti, C., Schiano, D.: Automatic Facial Expression interpretation: Where human-computer interaction, artificial intelligence and cognitive science intersect. Pragmatics and Cognition 8(1), 185–235 (2000)

15. Goldstein, E.: Sensation and Perception, 6th edn., ch. 13. Pacific Grove Wadsworth Publishing Company (2002)
16. Sherrick, C., Cholewaik, R.: Cutaneous Sensitivity. In: Willy (ed.) Handbook of Perception and Human Performance, vol. II. Cognitive Processes and Performance (1986)
17. Geldard, F.: Some Neglected Possibilities of Communication. Science 131, 1583–1588 (1960)
18. Verrillo, R.: Investigation of Some Parameters of the Cutaneous Threshold for Vibration. Journal of Acoustical Society of America, 1768–1773 (1962)
19. Verrillo, R.: Temporal Summation in Vibrotactile Sensitivity. Journal of Acoustical Society of America, 843–846 (1965)
20. Doren, C.V.: The Effects of a Surround on Vibrotactile Thresholds: Evidence for Spatial and Temporal Independence in the non-Pacinian I Channel. Journal of Acoustical Society of America, 2655–2661 (1990)
21. Kaczmarek, K., Bach-Y-Rita, P.: Tactile Displays. In: Barfield, W., Furness, T.A. (eds.) Virtual Enviornments and Advanced Interface Design. Oxford, New York (1995)
22. van Erp, J.: Guidelines for the Use of Vibro-Tactile Displays in Human Computer Interaction. In: Proceedings of Eurohaptics 2002, pp. 18–22 (2002)
23. Craig, J.: Difference Threshold for Intensity of Tactile Ttimuli. Perception and Psychophysics, 150–152 (1972)
24. Geldard, F.: Cutaneous Stimuli, Vibratory and Saltatory. Journal of Investigative Dermatology, 83–87 (1977)
25. Goff, D.: Differential Discrimination of Frequency of Cutaneous Cechanical Vibration. Journal of Experimental Psychology, 294–299 (1968)
26. Kaaresoja, T., Linjama, J.: Perception of Short Tactile Pulses Generated by a Vibration Motor in a Mobile Phone. In: Proceedings of World Haptics 2005, pp. 471–472 (2005)
27. Karlsson, J., Eriksson, J., Li, H.: Real-time Video over Wireless Ad-hoc Networks. In: Proceedings of the fourteenth International Conference on Computer Communications and Networks, ICCCN 2005 (2005)
28. Jones, C., Sivalingam, K., Agrawal, P., Chen, J.: A Survey of Energy Efficient Network Protocols for Wireless Networks. Wireless Networks 7, 343–358 (2001)
29. Jung, J., Choi, S.: Power Consumption and Perceived Magnitude of Vibration Feedback in Mobile Devices. In: Jacko, J.A. (ed.) HCI 2007. LNCS, vol. 4551, pp. 354–363. Springer, Heidelberg (2007)
30. Yanagida, Y., Kakita, M., Lindeman, R., Kume, Y., Tetsutani, N.: Vibrotactile Letter Reading using a Low-resolution Tactor Array. In: Proceedings of 12th International Symposium on Haptic Interfaces for Virtual Environment and Teleoperator Systems, pp. 400–406 (2004)
31. Tain, Y., Kanade, T., Chon, J.: Recognizing Action Units for Facial Expression Analysis. IEEE Tran. Pattern Analysis and Mach. Intelligence 23, 97–115 (2001)
32. Ekman, P., Friesen, W.: Facial Action Coding System: A technique for the measurement of facial movements. Consulting Psychologist Press, Palo Alto (1978)
33. Shi, J., Tomasi, C.: Good Features to Track. In: IEEE Conf. on Computer Vision and Pattern Recognition, CVPR 1994 (1994)
34. Réhman, S.U., Liu, L., Li, H.: Lipless Tracking and Emotion Estimation. In: IEEE 3rd int. Conf. on Signal-Image Technology & Internet-based Systems (2007)
35. Matthews, I., Chootes, T., Bangham, J., Cox, S., Harvey, R.: Extraction of Visual Features for Lipreading. IEEE Tran. Pattern Analysis and Mach. Intelligence 24, 198–213 (2002)

36. Kass, M., Witkin, A., Terzopoulos, D.: Snakes: Active Contour Models. International Journal of Computer Vision 1, 321–331 (1987)
37. Faulkner, X.: Usability Engineering. Palgrave Macmillan, Basingstoke (2002)
38. Devore, J.L.: Probability and Statistic for Engineering and the Science. Brooks/Cole Publishing Company (1999)
39. Humanware: HP iPAQ 2490b,
    http://www.humanware.com/en-europe/products/blindness/
    handheld_computers/_details/id_28/maestro.html
40. Google: Google Eye-Free, http://code.google.com/p/eyes-free/
41. CASBLiP: Cognitive Aid System for Blind People,
    http://casblipdif.webs.upv.es/
42. Toledo, D., Magal, T., Morillas, S., Peris-Fajarnés, G.: 3D Environment Representation through Acoustic Images. Auditory Learning in Multimedia Systems. In: Proceedings of current Developments in technology-assisted education, pp. 735–740 (2006)
43. Loomis, J., Lederman, S.: Tactual Perception. In: Boff, K., Kaufman, L., Thomas, J. (eds.) Handbook of Perception and Human Performance, New York (1986)
44. Gescheider, G.A.: Psychophysics: The Fundamentals, 3rd edn., ch. 9-11. Lawrence Erlbaum Associates Publishers, Mahwah (1997)
45. Hayward, V., Astley, O., Cruz-Hernandez, M., Grant, D., Robles-De-La-Torre, G.: Haptic Interfaces and Devices. Sensor Review 24, 16–29 (2004)
46. Laycock, S., Day, A.: Recent Developments and Applications of Haptic Devices. Computer Graphics Forum 22, 117–132 (2003)
47. Otaduy, M., Lin, M.: Introduction to Haptic Rendering. In: ACM SIGGRAPH Course Notes on Recent Advances in Haptic Rendering & Applications, pp. 3–33 (2005)
48. Poupyrev, I., Maruyama, S., Rekimoto, J.: Ambient Touch: Designing Tactile Interfaces for Handheld Devices. In: Proceedings of the 15th annual ACM symposium on User interface software and technology, pp. 51–60 (2002)

Mobile Visual Aid Tools for Users with Visual Impairments

Xu Liu[1], David Doermann[2], and Huiping Li[2]

[1] Ricoh Innovations, Inc.,
Menlo Park, CA
liu@rii.ricoh.com
[2] Applied Media Analysis Inc.
387 Technology Drive
College Park, MD 20742
{doermann,huiping}@mobileama.com
http://www.mobileama.com

Abstract. In this chapter we describe "MobileEye", a software suite which converts a camera enabled mobile device into a multi-function vision tool that can assist the visually impaired in their daily activities. MobileEye consists of four subsystems, each customized for a specific type of visual disabilities: A color channel mapper which can tell the visually impaired different colors; a software based magnifier which provides image magnification as well as enhancement; a pattern recognizer which can read currencies; and a document retriever which allows access to printed materials. We developed cutting edge computer vision and image processing technologies, and tackled the challenges of implementing them on mobile devices with limited computational resources and low image quality. The system minimizes keyboard operation for the usability of users with visual impairments. Currently the software suite runs on Symbian and Windows Mobile handsets. In this chapter we provides a high level overview of the system, and then discuss the pattern recognizer in detail. The challenge is how to build a real-time recognition system on mobile devices and we present our detailed solutions.

Keywords: Visual Impairments, Camera Phone, Image Processing.

1 Introduction

Visual impairments affect a large percentage of population in various ways [1] including color-deficiency, presbyopia, low vision, and other more severe visual disabilities. Current estimates suggest there are approximately 10 million blind or visually impaired individuals in the United States alone. Visual impairment significantly affects quality of life of these populations and most have no effective cure. Devices that provide visual enhancement are in large demand, but are often expensive, bulky and dedicated to a single task. For example, an optical image enhancer such as magnifier can provide basic zoom function but cannot enhance the contrast, brightness, color and other details of the image. Electronic image

X. Jiang, M.Y. Ma, and C.W. Chen (Eds.): WMMP 2008, LNCS 5960, pp. 21–36, 2010.
© Springer-Verlag Berlin Heidelberg 2010

enhancers ("Acrobat LCD" and "Sense View Duo" for example) are powered by digital image processing technology and these programmable devices can be customized for various requirements of the visually impaired users. This device, like others, is a special purpose piece of hardware.

We are exploring a unique opportunity to use a phone's camera as an electronic eye (a MobileEye) to help users with visual impairements see and understand their surroundings. With increases in popularity in recent years, camera phones have become a pervasive device available to a large percentage of individuals[2,3]. The combined imaging, computation and communication capabilities have inspired us to apply image processing and computer vision technologies on the device in new and exciting ways.

Compared with optical and electronic image enhancement tools, a camera phone has some unique advantages. First, it is a portable hand-held device which is already carried by a large number of people. Second, new functions can be easily programmed and customized in software without extra hardware. Third, the communication capability of these devices opens a wide range of opportunities to provide access to knowledge and computational resources from the internet. Utilizing these advantages of the camera phone, we have built the MobileEye system which consists of four major subsystems:

1. A color mapper that helps the color blind user distinguish colors.

2. A software magnifier that enhances the detail and contrast of an image to facilitate reading and understanding.

3. A camera based document retrieval system which finds the document from a large document database using only a snapshot of the page and speaks it out using the text-to-speech (TTS) techonology.

4. A pattern recognizer that recognizes certain objects such as currencies as the camera phone approaches the object.

2   Overall Design Principles

Since the MobileEye system is designed for people with visual impairments, it is impractical for our user to complete sophisticated operations which are already difficult for regular users because of the small keyboard input [4]. The overall design principle we follow is to minimize user operation, especially with respect to keyboard hits. Take the pattern recognizer as an example. A typical mobile pattern recognition system (for example the Kurzweil K1000 device) asks the user to take a snapshot and then the system tries to recognize the result. If the image is imperfect, the recognition may fail and the user will have to repeat the process. However we cannot expect visually impaired user to perform such tasks and it is impractical to ask them to take high quality pictures for recognition. We choose to process the image in real time which provides a much smoother user experience. Our pattern recognizer runs on the device and processes approximately 10 frames per second so that the user gets instant response as the camera moves. This introduces the challenge that we must process the video stream from the camera at a very high speed. We address this problem using a boosted object detector with both high efficiency and accuracy.

3   MobileEye Tools

In this section we briefly introduce three sub-systems of the MobileEye tools and leave the pattern recognizer in the next section for detailed discussion.

3.1   Color Mapper

A color vision deficiency is sometimes referred to as color blindness. In fact it is very rare that one is completely "blind" to a color, but deficiency such as red-green color blindness is as high as 8% in Caucasian, 5% in Asian and 4% in African males. Color blind is not a severe visual impairment and is often ignored by visual appearance design. A typical example of poor design to red-green color blind is show in Figure 1. Such graphics and figures are misleading to people with color deficiency but widely exist on web sites, slides and posters. It is not easy to correct them using optical devices but digital imaging devices fit this task. Our color channel mapper processes a camera captured image in real-time by swapping color channels. Depending on the type of color deficiency, we take one of the two most confusing color channels and swap it for a third color channel. For example, the red channel is swapped with blue for red-green color blind vision. Figure 2 shows how the image is processed.

3.2   Software Magnifier

Low vision and Presbyopia are quite common in the elderly population. Most people wear corrective lens or carry a magnifier to help reading. Our software magnifier performs like a glass magnifier but has three distinct advantages. First, the magnification is actually a digital zoom and is not limited or fixed. Text can be zoomed as long as it fits in the screen. Second, the contrast can be enhanced at the same time so that text is distinguishable from background to facilitate

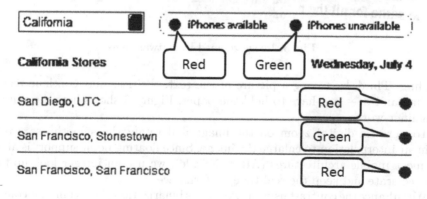

Fig. 1. A color design without considering the color blind user from www.apple.com

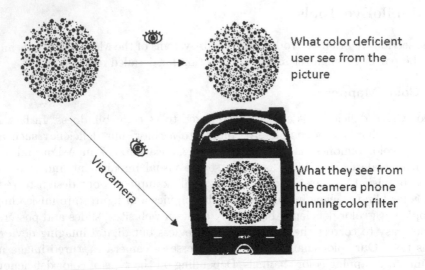

What color deficient user see from the picture

Via camera

What they see from the camera phone running color filter

Fig. 2. Color mapper

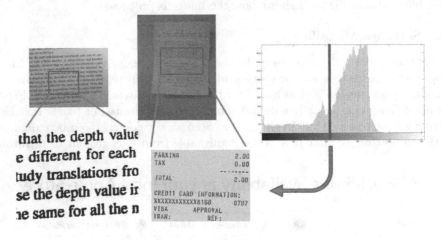

that the depth value
e different for each
tudy translations fro
se the depth value ir
ne same for all the n

```
PARKING              2.00
TAX                  0.00
                   --------
TOTAL                2.00

CREDIT CARD INFORMATION:
XXXXXXXXXXXX8168        0707
VISA        APPROVAL
TRAN:          REF:
```

Fig. 3. Image zoomed by software lens

reading. Third, by taking a picture of the text, the user may continue reading off line and does not have to hold the paper. Figure 3 shows how the software magnifier works.

To perform digital zoom on the image and retain the smoothness, we use bilinear interpolation to enlarge the image. Since floating point support is absent on most mobile architectures (ARM, XScale), we use an integer look-up-table to accelerate and keep the real-time performance.

We enhance the contrast using a two means binarization algorithm that converts the image into black and white. Two means binarization adapt to lighting and can distinguish foreground from background by choosing the threshold at the middle of two peaks from the gray scale histogram of the image. Other types of image enhancement such as edge enhancement can also be added programmatically.

3.3 Document Retriever

Although the software magnifier can aid in reading text, it is difficult to read large quantities of text on a small screen. To address this problem, we have designed a camera based document retriever and incorporated the text to speech technology. Our document retriever system is based on the observation that most of today's documents (newspapers, magazines etc.) have both paper version and electronic version that can be accessed. J.J.Hull et al.[5] have proposed a novel paper based augmented reality (PBAR) system that links small patches of physical documents to online contents. PBAR system runs in real time on a small scale database (250 documents) so that the user may click on the dynamic hyperlinks as the camera moves across the document. Our system runs off line but on a larger scale (100k pages) document database.

The data flow of mobile retriever is shown in Figure 4. The user takes a snapshot of a page and the pre-installed software sends the image to the server as a query. If the image content matches any document in the database, the article will be read back via the phone speaker. Sophisticated retrieval and TTS algorithms are performed on the sever side, but all the user has to do is to capture a small, low resolution portion of the document.

We do not expect our user to take a perfect full page snapshot. We extract unique features from the partial document image and use their geometrical relationship to verify our retrieval. Our technology does not rely on OCR (Optical

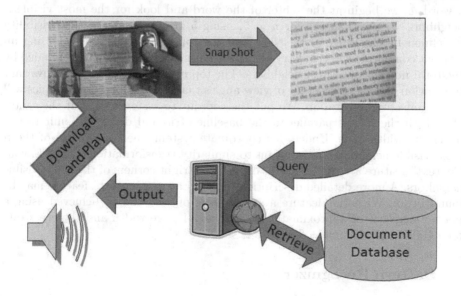

Fig. 4. Dataflow of document retriever

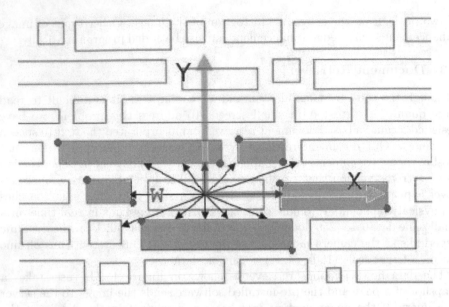

Fig. 5. Layout Context feature for document retriever

Character Recognition) so a perfect document image is not required. We use the "Layout Context" feature to uniquely identify document patches.

The "Layout Context" features are extracted from the geometrical location of the words' bounding boxes of the document patch. As shown in Figure 5 each rectangle shows a bounding box of a word. To extract the "Layout Context" of a word w, we begin at the center of the word and look for the most visible n neighbors. Figure 5 shows, for $n = 5$, using 5 green rectangles. The visibility is defined by the angle of the view and the top n can be extracted using an efficient computational geometry algorithm with complexity linearly bounded by the total number of nearest neighbors. The top n visible neighbors are invariant to rotation and the percentage of view angles that a neighbor word occupies will not be effected by rotation. We put the coordinate system origin at the center of w with the X-axis parallel to the baseline of w and define the unit metric using the width of w. Under this coordinate system, the coordinates of the n most visible neighbors are invariant to similarity transformations. The "Layout Context" feature are the L2-norms of the lower right corners of the top n visible neighbors. A more detailed description of the "Layout Context" feature may be found in [6]. With this feature a success rate of 96% can be achieved using a partial snapshot of approximately 1/3 of an A4 page width and the error rate on a 100K page document database is < 1%.

4   Pattern Recognizer

Pattern recognition has been developed over the past several decades to use sophisticated machine learning and classification algorithms that can achieve very high accuracy. For selected applications, these techniques however usually run slow which does not fit the profile of a resource constrained mobile device. It is even harder when the pattern recognizer has to run in real time as stated in our design principle. We have designed a general pattern recognition framework for the mobile device using the Adaboost[7] algorithm and target a specific task - to recognize U.S. currencies. U.S. currency is not designed to be easily recognized by the visually impaired population because different face values can only be distinguished visually. The blind community has recently won the law suit against Department of Treasury [8] for discrimination because it has failed to design and issue paper currency that is readily distinguishable by the blind. We use the camera phone to recognize bills[1] and communicate the value via a speaker or vibration.

4.1 Related Techniques

Classic pattern recognition algorithms usually include feature extraction and feature classification as two core components. Widely used features such as SIFT[9] or SIFT-like[10] have high repeatability. SVM [11] and neural networks [12] can be trained to achieve high accuracy given enough time and space allowance. However, these classic pattern recognition approaches cannot be ported directly to mobile devices. As we noted in the introduction, implementing pattern recognition on mobile devices has three major challenges. 1) The limited processing power of the device, 2) the fact that the captured scene could contain complex background resulting in false positive that must be eliminated, and 3) the expectation of the user who typically expects instant feedback and requires online (real time) recognition.

These three challenges are related to the speed and efficiency of the algorithm. The algorithm must be efficient enough to fit in the light-weight device and be able to discard images quickly or pixels that are not of interest so more time can be allocated to the image that contains objects to be recognized. Ideally, when an algorithm is efficient enough that it can run in real time, the recognition can be performed on the video stream of the camera, and the user does not have to hit a key to capture an image. Real time recognition gives a smooth user experience and avoids motion blur caused by "click-to-capture," but, as noted earlier, it must typically deal with lower quality data.

4.2 Fast Classification with Random Pixel Pairs

In this section we introduce a fast classifier based on random pixel pairs. One challenge of pattern recognition on the mobile device is usability. Traditional "capture and recognize" approaches may not be friendly to mobile users. When the recognition fails (because of motion blur, de-focusing, shadows, or any other reason) the user will have to click and try again. Instead, real time recognition

---

[1] Our currency reader was awarded the first place of the ACM Student Research Competition 2009 (http://www.acm.org/src/)

Fig. 6. Currency reader for the visually impaired

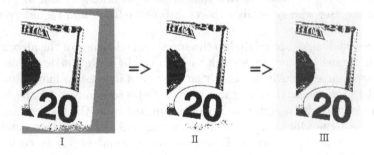

Fig. 7. Background subtraction and feature area extraction

is preferred. At the same time, recognition may benefit from the fact that images are taken by a cooperative user. Unlike a mounted camera with the objects moving in the scene, the camera (phone) is moving itself and the user is usually approaching the object to be recognized. We assume that the object rests approximately to the same relative position of the camera when being recognized. Under this assumption, we can subtract the background, extract and normalize the object, then perform the recognition under a relative stable setup. Our primary concern, the features, is the key subject of this section. The features we seek must be distinct (not to raise any false alarms), robust (to tolerate weak

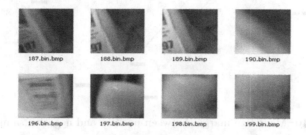

Fig. 8. Twenty dollar bills

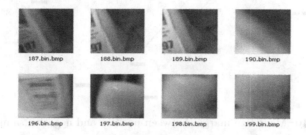

Fig. 9. Negative samples: arbitrary scene

perspective and lighting variation) and fast to compute on the phone. When considering feature, the first option may be SIFT [9] key points which outperform most of other features in terms of accuracy. However, the speed of SIFT is a significant challenge for mobile devices. The Gaussian convolutions are too computationally intensive for most of today's mobile devices. Wagner et al. [13] has presented a very promising modified SIFT algorithm. In their approach the SIFT feature vector was computed at a fixed scale, the scale-invariance was traded for speed. The modified SIFT fits perfectly for pose estimation but may not be directly applicable for object recognition. Meanwhile some recent results of efficient and robust feature (point) extraction are built from simple elements such as box filters and random pixel pairs [14,15].

Our feature set consists of a large set of binary values of pairwise intensity comparisons. For a $w \times h$ image, there are as many as $C^2_{w \times h}$ different pairs of pixels to compare. Our goal aims to find those pairs that uniquely define the object of interest. We achieve this goal with a learning approach.

Consider attempting to recognize a twenty dollar bill (Figure 8). We first perform background subtraction and normalization, and collect a set of positive samples (Figure 8) and an even larger set of negative samples, Figure 9. We train our recognizer by selecting discriminating intensity pairs. For each pair of pixels (i, j), we define $P^+(i, j)$ to be the number of positive samples with greater intensity at pixel i than at pixel j, similarly, define $P^-(i, j)$ to be the number of negative samples with greater intensity at pixel i than at pixel j. Our goal will then be to choose pairs (i, j) to maximize $P^+(i, j)$ and minimize $P^-(i, j)$. One

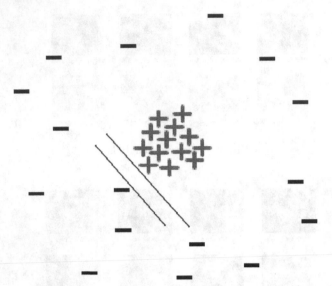

Fig. 10. Maximize margin between positive and negative samples

naive way to achieving this goal is to maximize $P^+(i,j)/P^-(i,j)$, but this will not work because the large collection of negative samples make $P^-(i,j)$ almost random. Nevertheless, numerous pairs satisfy all the positive samples, among which we would like to choose the most distinct ones.

Although the number of hits of a pair (i,j) in the negative samples cannot help us judge whether the choice of (i,j) is good, it can help measure the distance from the closest negative sample to the positive samples. As shown in Figure 10, we will maximize the margin between positive samples and the closest negative samples by scoring positive and negative samples. A higher score indicates a higher probability of an inlier and lower scores for outliers. After training, we can use a threshold on scores to classify the pattern. Using this criteria, we develop the following algorithm:

Assign an initial score of zero for all negative samples and keep a pointer m that always points to the highest score. This can be done efficiently using a heap.

1. Generate a random pair (i,j).
2. If (i,j) satisfies negative sample m, then go back to step 1.
3. If (i,j) does not satisfy the all positive samples, go back to step 1.
4. For all negative samples, increase its score by 1 if it satisfies pair (i,j) and modify pointer m to point to the negative sample with highest score (hence, the score of m is not increased).
5. Go back to 1 until we have n pairs.

Using this algorithm, we collect n discriminating pairs that represent the object. Suppose we also have a score for the positive samples. During each round, all positive samples get increased by 1, while the closet negative sample does not.

Fig. 11. The normalized feature area and random pixel pair

The gap between positive and negative samples is enlarged. In the recognition phase, we will use this score to judge whether an object is recognized or rejected, and our threshold will be placed in the middle of the gap. Ideally we would like to see all positive samples with high scores and all negative samples with low scores. However, there might be a negative sample that is similar to the inliers, and our algorithm can spot the most confusing outliers and establish a boundary between the inliers and those outliers. The accuracy of this detection algorithm is further enhanced using the Ada-boost[7] algorithm.

4.3   Initial Design

In order to detect and recognize a bill, we first binarize the image and remove irrelevant background. Black pixels touching the boundary (Figure 7-I) of the image are regarded as backgrounds since the bill always has a white boundary along the edge. After removing the background some noise (Figure 7-II) might still exist. We further refine the location of a bill by running a breadth-first-search (BFS) from the image's center to remove the remaining noise. The complexity of this step is linear in the number of pixels in the image and after processing we know the exact position of the feature area (Figure 7-III). We then normalize extracted area to a rectangle with an aspect ratio of 4:1 (Figure 11) for recognition.

We collected 1000 samples of captured currencies of each side of the most common U.S. bills. Each has four sides, two front and two back. We also collected 10000 samples of general scenes which are not currency. For each side of a given bill, we use Ada-boost[7] to train a strong classifier from a set of weak classifiers. The weak classifiers must be computationally efficient because hundreds of them must be computed in less than 0.1 second.

We define a weak classifier using 32 random pairs of pixels in the image. A random pair of pixels have a relatively stable relationship in that one pixel is brighter than the other. An example of a random pair is shown in Figure 11 where pixel A is brighter than pixel B. The advantage of using pixel pairs is that their relative brightness is not affected by environmental lighting variations. Since the same relationship may also occur in general scenes, we select the pairs that appear more frequently in the inliers (currency images) and less frequently in the outliers (non-currency images).

A weak classifier will provide a positive result if more than 20 pairs are satisfied and negative otherwise. 10 weak classifiers selected based on Ada-boost form a strong classifier that identifies a bill as long as it appears in the image. To recognize a bill we only need $32 \times 10 = 320$ pair-wise comparisons of pixels. Our system is trained to read \$1,\$5,\$10,\$20,\$50 and \$100 bills and can process 10 frames/second on a Windows Mobile (iMate Jamin) phone at a false positive rate $< 10^{-4}$.

It should be pointed out that this framework is general so that new notes (e.g. \$2) can be easily added to the system. To demonstrate the extendability of our system and the potential to help the visually impaired in other aspects, we also trained it to recognize 16 logos.

4.4   Revised Design

Although the initial design of the currency reader satisfies our primary require-ments of real time recognition and has a high accuracy, it could be further improved after an experimental study of its practical use. Users with visual dis-abilities identified two major disadvantages of the initial design. First, it required the coverage of the entire right hand side of the bill, i.e., the upper right and bottom right side of the bill must be captured at the same time. However, it may be difficult to accomplish such coverage without a great deal of practice. Second, users with visual disabilities liked to fold the bills in different ways to distinguish among denominations, but folding can change the shape of the right hand side of a bill and may disturb the recognition.

This suggests the use of a smaller feature area for recognition because it is easier to capture and less likely to be disturbed by folding. We have refined our currency reader to identify a feature area with the number denomination as shown in Figure 13. Feature areas are first detected using a fast pre-classifier and then identified using a strong classifier based on random local pixel pairs, as described in Section 4.2.

In our approach, we tackle these changes in two steps to achieve high accuracy, real time recognition on mobile devices. First, we use a very fast pre-classifier to filter the images and areas of an image with low probability of containing the target. This step is inspired by the Viola-Jones face detector [16] and the Speed-Up Robust Feature [15], both of which use box filters to detect objects and features rapidly. Second, we use a set of local pixel pairs to form weak classifiers and use ada-boost [7] to train strong but fast classifiers from these weak classifiers. The idea of using local pixel pairs is inspired by Ojala, et al.'s work on Local Binary Patterns (LBP[17]) and more recent work by Pascal, et al. [14]. The advantage of local pixel pairs lies in their robustness to noise, blur and global lighting changes which are significant challenges for mobile recognition. The details are presented in the next two sub-sections.

4.5   Pre-classification and Filtering

To detect an object in an image, an exhaustive search is usually inefficient be-cause most of the areas in the image do not contain the object in which we are

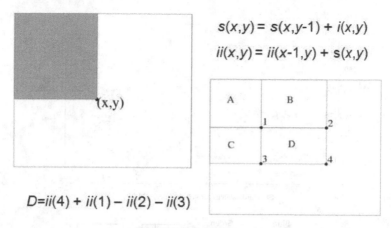

Fig. 12. Integral image

interested. A pre-classification which filters these areas is therefore important and can speed the detection by ten times or more. In our research we found that a box filter computed using an integral image is very efficient and can be applied to mobile devices. In an integral image, at each pixel the value is the sum of all pixels above and to the left of the current position. The sum of the pixels within any rectangle can be computed in four table lookup operations on the integral image in constant time as shown in Figure 12. If we replace the original image with an image with the squared gray scale value at each pixel, we can then compute the standard deviation (second order moment) within any rectangle in $O(1)$ time. Any order of moment can be computed in $O(1)$ time using an integral image. Both the Viola-Jones face detector[16] and SURF[15] benefit from the speed of the box filter and integral image.

We found that the standard deviation (STD) in the sub-image of an object is relatively stable and combination of STDs can be used as a cue to filter non-interest image areas. In Figure 13, we divide the feature area of a twenty dollar bill into 6 boxes (3 vertical and 3 horizontal). The STD in each sub-window falls in a relatively stable range and we search only within these ranges for the potential corner patterns to recognize. In each sub-window an STD range may span at most $1/2$ (red) or even less (blue) of possible STD. Assuming the STD in each sub-window is independent and is equally distributed in an arbitrary scene, the box filter can eliminate $1 - (1/2)^6 = 98.4\%$ of the computation by discarding low probability regions. In our experiment we found the pre-classification can speed the algorithm by 20 times on a camera phone.

4.6 User Interface and Evaluation

To meet the requirements of users with visual disabilities, we pay special attention to the details of the user interface. Every operation of the software is guided by a voice message. It requires two key presses to activate the camera to prevent accidental activation. The software automatically exits after being idle for two

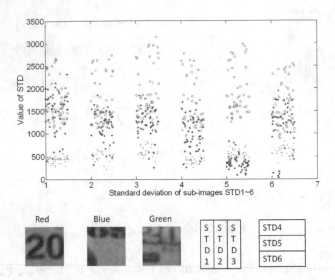

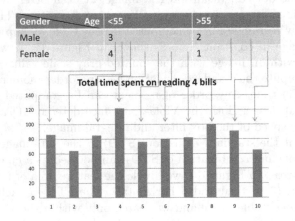

Fig. 13. Standard deviation of 6 sub-images at 3 corners of a 20 dollar bill

Fig. 14. User evaluation of mobile currency reader

minutes to save battery power. The user has the option of "force" recognition of a bill by pressing the center button. The software will search for additional scales and positions for the feature area in "forced" recognition.

We have performed user evaluation with the system of the refined design. Ten blind users were asked to identify four bills using the camera phone currency reader. Each user was given a brief two minute introduction on how to use the device and the software, they were asked to continue until all four bills are recognized. The total time (including entering and exiting the program) was recorded to measure the usability of the software. On average, users recognized a bill in 21.3 seconds, as shown in Figure 14. Our currency reader raised no false positive during the experiment.

5 Conclusion

We have introduced our MobileEye system which aids individuals with visual impairments to better see and understand their surroundings. Our system requires little effort to operate and can be deployed to a wide varieties of camera enabled mobile devices. We have tested the usability of our solution with visually impaired users and improved our algorithm. The major concern of our future research is the time of responding e.g. how long it will take a user to recognize a bill using the device running our software, how quick can a document be successfully retrieved. We will also test if they feel comfortable with the way that the result is communicated via TTS or vibration. We do face one major challenge in reaching the end user. Most handsets together with their software applications are deployed by the wireless service providers. We have put our Color Mapper subsystem online and received 140 downloads and some positive feedback. Our end user with visual disabilities, however, may not have the knowledge and skill to download and install the software themselves. It may require the cooperation of service providers and probably government support to promote the MobilcEyc system to a larger number of users.

References

1. Massof, R.W., Hsu, C.T., Barnett, G.D., Baker, F.H.: Visual Disability Variables. I,II: The Importance and Difficulty of Activity Goals for a Sample of Low-Vision Patients. Archives of Physical Medicine and Rehabilitation 86(5), 946–953 (2005)
2. Milanesi, C., Zimmermann, A., Shen, S.: Forecast: Camera phones, worldwide, 2004-2010. Gartner Inc. Report No. G00144253 (2006)
3. Kindberg, T., Spasojevic, M., Fleck, R., Sellen, A.: The ubiquitous camera: An in-depth study of camera phone use. IEEE Pervasive Computing 4(2), 42–50 (2005)
4. Karlson, A., Bederson, B., Contreras-Vidal, J.: Understanding One-Handed Use of Mobile Devices. In: Handbook of Research on User Interface Design and Evaluation for Mobile Technology (2008)
5. Hull, J., Erol, B., Graham, J., Ke, Q., Kishi, H., Moraleda, J., Van Olst, D.: Paper-Based Augmented Reality. In: 17th International Conference on Artificial Reality and Telexistence, pp. 205–209 (2007)
6. Liu, X., Doermann, D.: Mobile Retriever: access to digital documents from their physical source. International Journal on Document Analysis and Recognition 11(1), 19–27 (2008)
7. Freund, Y., Schapire, R.: A Decision-Theoretic Generalization of On-Line Learning and an Application to Boosting. Journal of Computer and System Sciences 55(1), 119–139 (1997)
8. http://www.acb.org/press-releases/final-edit-paper-currency-ruling-080520.html
9. Lowe, D.: Distinctive image features from scale-invariant keypoints. International Journal of Computer Vision 60(2), 91–110 (2004)

10. Fergus, R., Perona, P., Zisserman, A.: Object class recognition by unsupervised scale-invariant learning. In: 2003 IEEE Computer Society Conference on Computer Vision and Pattern Recognition, vol. 2 (2003)
11. Cortes, C., Vapnik, V.: Support-vector networks. Machine Learning 20(3), 273–297 (1995)
12. Specht, D.: Probabilistic neural networks. Neural Networks 3(1), 109–118 (1990)
13. Wagner, D., Reitmayr, G., Mulloni, A., Drummond, T., Schmalstieg, D.: Pose tracking from natural features on mobile phones. In: 7th IEEE/ACM International Symposium on Mixed and Augmented Reality, ISMAR 2008, pp. 125–134 (2008)
14. Özuysal, M., Fua, P., Lepetit, V.: Fast keypoint recognition in ten lines of code. In: IEEE Conference on Computer Vision and Pattern Recognition, CVPR 2007, pp. 1–8 (2007)
15. Bay, H., Tuytelaars, T., Van Gool, L.: SURF: Speeded up robust features. In: Leonardis, A., Bischof, H., Pinz, A. (eds.) ECCV 2006. LNCS, vol. 3951, pp. 404–417. Springer, Heidelberg (2006)
16. Viola, P., Jones, M.: Robust real-time face detection. International Journal on Computer Vision 57(2), 137–154 (2004)
17. Ojala, T., Pietikainen, M., Maenpaa, T.: Multiresolution gray-scale and rotation invariant texture classification with local binary patterns. IEEE Transactions on Pattern Analysis and Machine Intelligence 24(7), 971–987 (2002)

QR Code and Augmented Reality-Supported Mobile English Learning System

Tsung-Yu Liu[1], Tan-Hsu Tan[2,*], and Yu-Ling Chu[2]

[1] Department of Multimedia and Game Science, Lunghwa University of Science and Technology
joye.liu@msa.hinet.net
[2] Department of Electrical Engineering, National Taipei University of Technology
thtan@ntut.edu.tw, chu_yuling@tp.edu.tw

Abstract. Mobile learning highly prioritizes the successful acquisition of context-aware contents from a learning server. A variant of 2D barcodes, the quick response (QR) code, which can be rapidly read using a PDA equipped with a camera and QR code reading software, is considered promising for context-aware applications. This work presents a novel QR code and handheld augmented reality (AR) supported mobile learning (m-learning) system: the handheld English language learning organization (HELLO). In the proposed English learning system, the linked information between context-aware materials and learning zones is defined in the QR codes. Each student follows the guide map displayed on the phone screen to visit learning zones and decrypt QR codes. The detected information is then sent to the learning server to request and receive context-aware learning material wirelessly. Additionally, a 3D animated virtual learning partner is embedded in the learning device based on AR technology, enabling students to complete their context-aware immersive learning. A case study and a survey conducted in a university demonstrate the effectiveness of the proposed m-learning system.

Keywords: Augmented Reality, Handheld Device, Immersive Learning, Task-based Learning.

1 Introduction

Globalization is a major index of a country's competitiveness. Given the leading role of English as an international language, the Taiwanese government has mandated numerous programs to strengthen the English language skills of students. However, among the factors that have limited the success of such programs are limited practice time that students have outside the classroom, lack of motivation in English learning activities, and absence of learning opportunities in actual circumstances. Therefore, in recent years, there has been emphasis on the application of information technology to resolve the abovementioned problems in English learning.

---

* Corresponding author.

X. Jiang, M.Y. Ma, and C.W. Chen (Eds.): WMMP 2008, LNCS 5960, pp. 37–52, 2010.
© Springer-Verlag Berlin Heidelberg 2010

In [1], the authors indicated that the mobility, flexibility, and instant access of handheld devices enable students to actively engage in highly interactive learning activities without constraints in time or location. The role of m-learning in improving language learning has also received considerable attention. For instance, in [2], the authors developed an adaptive computer-assisted language learning software for mobile devices called Mobile Adaptive CALL (MAC). MAC helps Japanese speakers of English in perceptually distinguishing between the non-native /r/ vs. /l/ English phonemic contrast to improve their discriminative capability. By adopting mobile computing and information technologies in [3], the authors developed a mobile-based interactive learning environment (MOBILE) to facilitate elementary school English learning. Their results demonstrated the effectiveness of MOBILE in enhancing students' learning motivation and learning outcomes. Although the aforementioned studies have effectively developed mobile English learning environments and activities to aid learning, there are rare studies on investigating the use of context-aware learning strategies in English learning. Context-aware systems featuring contextual data retrieval, engaging learning experiences, and improved learning effects are described in [4]. Thus, it is worth investigating how a context-aware m-learning environment benefits English learning.

This work presents a context-aware m-learning environment called handheld English language learning organization (HELLO) that provides interesting learning activities to increase students' motivation in English learning. Students in this learning environment actively engage in English learning activities without constraints in time or location, thus upgrading their English language skills. Additionally, a case study conducted on a university campus demonstrates the effectiveness of the proposed English learning environment. This work has the following objectives:

- To develop a 2D barcode and an augmented reality-supported mobile English learning environment that enables situated and immersive learning;
- To develop collaborative, situated, immersive, and m-learning activities by applying the proposed learning environment in order to improve students' learning interest, motivation, and outcomes; and
- To understand how the proposed learning environment and its related learning model influence student attitudes toward learning as well as to assess the degree of system acceptance by administering a questionnaire survey.

2 Literature Review

Recent advances in wireless communication technologies have led to the evolution of an m-learning model. Mobile learning is superior to e-learning in terms of flexibility, cost, compactness, and user-friendliness [5]. With the assistance of wireless technologies and handheld devices, an m-learning environment can be easily created to facilitate the objectives of learning without time and location constraints as well as in various formats, which are impossible in traditional classroom learning.

Features of contextual data retrieval, active engagement in learning, and enhanced learning outcomes that are characteristic of context-aware systems have been extensively adopted in various learning activities [4]. In [6], the authors coined the term "context-aware," in which context is regarded in terms of location, identities of nearby individuals and objects, and subsequent changes to those individuals and objects. In [7], the author defined "context" as contextual information that can characterize an entity that can be an individual, location, or physical object that is viewed as relevant to the interaction between a user and an application. Several studies have developed various context-aware learning systems to improve language learning. For instance, in [8], the authors developed a tag added learning objects (TANGO) system, capable of detecting objects around learners and providing learners with object-related language learning materials by radio-frequency identification (RFID) technology.

Augmented reality (AR) is highly promising for integration in an m-learning environment for improving learning outcome and learning experience by immersion. Immersive learning allows individuals to experience feelings and emotions as they do in the real world by interacting in a virtual environment. Many studies have developed AR-based learning systems to enhance immersive learning. For instance, in [9], the authors developed a wearable AR learning system, namely, MagicBook, in which a real book is used to seamlessly transport users between reality and virtuality. That work adopted a vision-based tracking method to overlay virtual models on real book pages, thereby creating an AR scene. Users seeing an AR scene and AR objects enjoy an immersive virtual reality (VR) world. Therefore, AR (or VR) is a valuable technology for students to acquire a richer learning experience and improve learning outcomes. Video cameras are normally embedded in mobile phones; thus, wireless local area networks (WLANs), Bluetooth, GSM (also known as Global System for Mobile Communication), and multimedia capabilities can assist students to learn without time and location constraints. To achieve the objectives of context-aware and immersive learning in English, this work presents a sensor technology and AR-supported context-aware m-learning environment with handheld phones for facilitating campus learning. In this environment, students are actively engaged in interesting English learning activities, thus enhancing their English language skills.

3 Implementation of a Mobile English Learning System

3.1 Implementation Issues

While context-aware m-learning provides a more situated and interactive learning experience than m-learning [10, 11], integrating situations into a context-aware m-learning environment poses a major challenge. Fortunately, advanced sensor technologies, including 2D barcode, Infrared Data Association (IrDA), global positioning system (GPS), Bluetooth, RFID, Zigbee, and WLAN can provide situated services. Table 1 compares various sensor technologies for positioning. Among these positioning technologies, 2D barcode technology is feasibly applied to mobile phones in context-aware m-learning.

Table 1. Comparison of positioning technologies

Characteristics	802.11	GPS	RFID	2D barcode
Positioning accuracy	Low	Low	High	High
Indoor	Yes	No	Yes	Yes
Context-awareness	Low	Low	High	Middle
Sensor technology	Auto	Auto	Auto	Passive
Cost	High	Low	High	Low
Cover Area	Micro	Wide	Micro	Micro
Practicability	Low	Low	High	High

2D barcode technology has many advantages, including a large storage capacity, high information density, strong encoding, strong error-correcting, high reliability, low cost, and ease of printing [12]. 2D barcode technology has thus become popular in various applications, including ticketing services, manufacturing, product identification, flow control, quality control, logistics management, interactive advertising, marketing, mobile commerce, business transactions, medical treatment, and location-based services.

2D barcode technology stores data along two dimensions, allowing it to contain a greater amount of information than a 1D barcode. Despite more than 200 2D barcode standards worldwide, only a few are widespread, including portable data file 417 (PDF417), data matrix, quick response (QR) code, and Magic Code. Of the 2D barcodes, QR code, as created by Denso-Wave in 1994, has become increasingly popular in Taiwan since QR code-decrypting software is embedded in many mobile phones. QR code requires only around 23 micro seconds for decoding; therefore, this work adopts it to assess user receptiveness to the proposed mobile learning system.

2D barcode software performs two basic functions of encoding and decoding. Table 2 lists established 2D barcode software providers. Developers can use these barcode toolkits to develop 2D barcode-based applications. Each product may only provide either a 2D barcode encoder or a decoder; alternatively, it either only supports Windows or Windows Mobile applications. A developer can use the software development kit (SDK) to develop diverse 2D barcode technology applications.

Moreover, AR is a highly effective educational application owing to its ability to embed digital objects into a real environment [13]. Creating an AR application involves superimposing virtual image on a live video. The AR tool has the following operation procedures: tracking a marker via a camera and then taking a series of snapshots regarding this marker in real time; decoding the internal code of the marker (which refers to a virtual image); and overlaying the virtual image on a live video.

To create AR applications, ARToolKit is one of the most widely used tracking libraries with more than 160,000 downloads. Developed by Kato in 1999 and subsequently released by the Human Interface Technology (HIT) Lab of University of

Table 2. 2D barcode software provider list

Platform/codec Provider	SDK for Windows		SDK for Mobile		
	encoder	decoder	encoder	decoder	free
Denso-Wave					
TEC-IT		√			
Lead Technologies	√	√	√	√	
Neodynamic	√		√		
Inlite Research, Inc.	√				
PartiTek, Inc.	√	√	√	√	√
AIPSYS.com	√	√	√	√	√
PyQrCodec	√	√			
MW6	√		√		
SIA DTK Software			√	√	
IDAutomation	√	√	√	√	
SimpleAct, Inc.					
iconlab Co., LTD.					
Yusuke Yanbe	√	√	√	√	√

Washington in [14], ARToolKit is maintained as an open source project hosted on SourceForge (http://artoolkit.sourceforge.net/) with commercial licenses available from ARToolWorks in [15].

ARToolKitPlus was developed internally as an integral part of the Handheld AR project in [16], later released to the public domain. ARToolKitPlus is an extended version of ARToolKit in [17]. ARToolKitPlus was succeeded by Studierstube Tracker, i.e., a computer vision library for detection and estimation of 2D fiducial markers. Studierstube Tracker was written with high performance for personal computers and mobile phones in [18]. Although ARToolKitPlus is available in the public domain, Studierstube Tracker requires a subscription fee. Several AR tool kits listed in Table 3 can be adopted to develop handheld AR applications.

Table 3. Argumented reality toolkit list

Organization	Toolkit	Platform	OS	SDK Fee
HIT Lab, University of Washington	ARToolKit	PC, laptop	XP, Mac OS, Linux	Free
Christian Doppler Lab , Graz University of Technology	ARToolKitPlus	PDA	PocketPC 2003 SE	Free
Christian Doppler Lab , Graz University of Technology	Studierstube Tracker	PC, mobile phone, PDA	XP, WinCE, Windows Mobile, Linux, Symbian, MacOS, iPhone	Charge
Augmented Environments Laboratory, Georgia Institute of Technology	OSGART	PC, laptop	XP, Vista, Mac OS, Linux	Free
University College London	MRT	PC, laptop	XP	Free

3.2 System Design

Fig. 1 illustrates the architecture of the proposed 2D barcode and AR-supported m-learning environment. The proposed learning environment consists of two subsystems: a HELLO server and m-Tools (application software). While teachers access the HELLO server through personal computers via the Internet, students communicate with the HELLO server from their mobile phones via WLAN. The functionalities of the two subsystems are as follows.

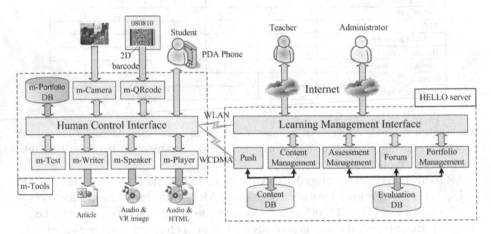

Fig. 1. Architecture of HELLO

The functionalities of the HELLO server are as follows:

- Content management unit (CMU): University administration assigns independent study courses and stores the learning materials in a content database (CDB).
- Assessment management unit (AMU): University instructors can give assessments to students to evaluate their learning outcome.
- Portfolio management unit (PMU): Students can upload their portfolios into an evaluation database (EDB) for review by instructor and grades evaluation via the PM unit.
- Forum unit: Through this unit, university instructors can instruct students to share their learning experiences with each other.
- Push unit (PU): Every day, this unit automatically delivers a sentence to students for daily practice.

A student with a PDA phone installed with m-Tools can learn English without location or time constraints. The functionalities of m-Tools are as follows:

- Listening and reading: The m-Player can download course materials and then students can read articles/news or listen to conversations from the HELLO server;

- Playing: The m-Player can play learning games or English songs;
- Speaking: To enhance speaking skills, students can use the m-Speaker. Students can practice speaking with the virtual learning tutor (VLT);
- Writing: Students can use the m-Writer to write an article or a diary entry in English;
- Context-awareness: When a student holds a PDA phone near the zone attached with 2D barcode technology, the m-Reader on that phone decrypts the internal code and sends it to the HELLO server. The HELLO server then downloads context-aware content to the PDA phone; and
- Evaluation: Students can use the m-Test to take tests and evaluate their learning achievements. Moreover, learning records can be stored in the m-Portfolio through the human control interface (HCI) after learning tasks are completed. Upon completion, the student learning portfolio can be uploaded into the EDB of the HELLO server for instructor review.

HELLO operates as follows. Teachers input materials and assessments into the CDB through the CMU, AMU, and PU. Teachers can review student portfolios and give grades through the PMU. The PU automatically delivers a daily English sentence to students' PDA phones via a wireless network, such as GSM and code division multiple access (CDMA), in order to enhance the listening skills of students.

Equipped with PDA phones to communicate with the HELLO server, students can access materials stored in a server via a WLAN. Students use m-Tools software to download articles, news, learning games, English comics, English songs, listening

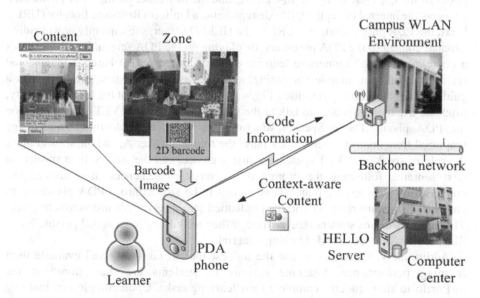

Fig. 2. Scenario of mobile English learning in the campus of the National Taipei University of Technology (NTUT)

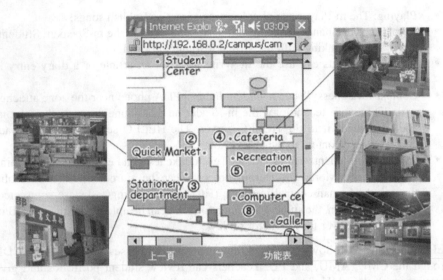

Fig. 3. Guide map of learning activity

materials, and conversational materials from the HELLO server, followed by use of the m-Player to play, listen to, and display learning materials. Additionally, each student holds a PDA phone near a zone that is attached to a 2D barcode. The student takes a photo of the QR code using the m-Camera, and the m-Reader on the PDA phone then decrypts the internal code; the QR code represents a Uniform Resource Locator (URL). Next, the PDA phone sends this URL to the HELLO server, subsequently downloading situated content to the PDA phone and displaying on the PDA screen. Fig. 2 illustrates a context-aware and immersive learning scenario based on 2D barcode, augmented reality, the Internet, mobile computing, and database technologies. Fig. 3 depicts a guide map of the learning activities. Fig. 4 presents an example of the learning activity. Students use the m-Speaker to talk to the virtual learning tutor (VLT) that appears on the PDA phone. The m-Speaker superimposes VLT on the learning zone image (captured from the m-Camera). VLT plays the role of speaker A, and the student plays the role of speaker B. VLT speaks the first sentence, and the student then speaks the next sentence following the prompt of conversation sentences in sequence. The conversation between VLT and the student can be stored into a PDA phone by an embedded software recorder and then uploaded into a server for instructors to grade. This features makes students feel as though they are talking to an actual person. Fig. 5 illustrates an example of AR learning material.

Additionally, students can use the m-Test tool to take tests and evaluate their learning performance. Learning records of students are then stored in the m-Portfolio after students complete their learning tasks. Upon completion, learning portfolios of students are uploaded into the EDB of the HELLO server for instructors to review.

Fig. 4. A scenario of learning activity: a student is talking with the virtual learning partner to practice conversation in a restaurant

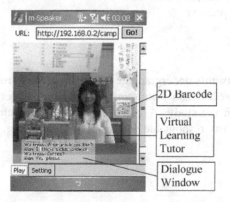

Fig. 5. An example of augmented reality learning material

4   Methodology, Course Design, and Experimental Procedure

A series of controlled experiments was performed with university students. Following completion of the experiments, a questionnaire was administered to students to evaluate the effectiveness of the HELLO in enhancing their learning motivation and learning outcomes.

4.1 Methodology

The questionnaire was administered to twenty students upon completion of the experiments (during the final class session) in order to determine the degree of perceived usefulness, user-friendliness, and attitudes toward the use of the HELLO server. A seven-point Likert scale was applied to all questions: 1 denoted strong disagreement, while 7 denoted strong agreement. The questionnaire results were analyzed using a one-sample t-test. The usefulness and user-friendliness of the system were evaluated using the technology acceptance model (TAM) [19, 20, 21]. TAM is an information system that creates models for how users accept and use a particular technology. TAM posits that two particular beliefs—perceived usefulness and perceived user-friendliness—are of priority concern. "Perceived usefulness" is defined as the subjective probability that the use of a given information system enhances a user's performance in an organizational context. "Perceived user-friendliness" refers to the degree to which the prospective user expects that using it is effortless. A user's "attitude toward using" is a function of the perceived usefulness and perceived user-friendliness that directly influences actual usage behavior [20, 22, 23, 24].

The internal consistency reliability of the questionnaire was evaluated using Cronbach's alpha coefficient. In [25], the author stated that 0.7 is an acceptable minimum reliability coefficient. Whether the two groups significantly differed from each other in terms of the pre-condition was determined using either an independent two-sample t-test or a pre-test.

4.2 Participants

Three instructors and twenty undergraduate freshmen selected from NTUT participated in the experiment. Two English instructors with teaching experience of more than a decade participated in the study. A computer science instructor with five years of teaching experience was responsible for installing, managing, and maintaining the computer system.

4.3 Course Design

Interaction and communication are essential to language learning [26, 27, 28]. Of the many communicative language learning approaches available, communicative language teaching (CLT) refers to language learning for the purpose of communicating. Additionally, task-based language learning (TBLL) focuses on asking students to complete meaningful tasks using the target language. Moreover, competency-based language teaching (CBLT) focuses on measurable and useable KSAs (knowledge, skills, and abilities). Furthermore, a natural approach (NA) focuses on "input" rather than practice. Among them, TBLL is the most effective pedagogical approach. In [29], the author stated that TBLL increases student conversations, relaxes the classroom atmosphere, and reinforces students' comprehensible input. Tasks refer to "activities where the target language is used by the learner for a communicative purpose in order to achieve an outcome" [30]. More than simply asking students to complete tasks sequentially, TBLL consists of three stages: pre-task, task cycle, and language focus [30]. Teachers discuss the topic with their classes, highlight useful words and phrases, help students understand task instructions, and prepare the pre-task stage. Each student group completes a common task collaboratively and then presents its findings to the class—or exchanges written reports—and compares the results during the task cycle stage. During the language focus stage, teachers help students practice new words, phrases, and patterns that occur in data either during or after the analysis. In [30], the author indicated that in TBLL, students can learn by doing. TBLL has the following characteristics: interactive, student-centered focused, meaningful materials, fluency language production, learning in the real world, and clear learning objectives [30, 31].

By helping students to collaborate with teachers and peers, a TBLL curriculum gradually helps students to use English meaningfully. By adopting the TBLL approach, this work designs a course entitled "My Student Life." Course topics include classrooms, libraries, a language center, gymnasiums, restaurants, dormitories, stadiums, cafeterias, the gallery, and the computer center. Mobile gamed-based learning, immersive learning, and context-aware learning are the pedagogical strategies. This course has the following learning objectives: to nurture listening, speaking and reading skills; to increase learning motivation through a designed learning game; and to enable students to learn in a real environment. Context-aware learning is then achieved through 2D barcode technology.

4.4 Procedures

A four-week experiment was performed with twenty undergraduate students, as follows. During the independent study phase (first two weeks), teachers introduced the HELLO system and demonstrated how to use the learning tools. A mobile task-based pedagogical strategy was adopted in the self-learning process. Students used PDA phones installed with m-Tools. A campus map appeared on the screen after students launched the game "My Student Life" on their PDA phones. The campus map had many zones marked on the map. Students simply clicked the desired zone, with the m-Player subsequently making available materials related to that zone. For instance, when a student selected the zone "Library," a library was displayed on the PDA phone. Students could also select the reading room to read an article, the newsroom to read news, or the multimedia room to watch a movie whenever desired. Importantly, students could learn without time or location constraints without going to an actual library.

During the context-aware learning phase (the other two weeks), students used the HELLO system to engage in the learning activity called "Campus Tour." Each student used a PDA phone installed with m-Tools and followed the guide map that appeared on the screen to engage in context-aware learning activities. When approaching a zone, a student used the PDA phone to take a picture and decrypt the 2D barcode. The detected identification code of the 2D barcode was then sent to the HELLO server via a WLAN. The HELLO server located the students and sent the context-aware contents back to their PDA phones. The VLT was superimposed with the zone image on the PDA screen. Next, students practiced conversation with the VLTs, similar to how they would talk with actual partners in the real world. Students visited the next zone after completing a conversation with VLT at a particular zone, until they had visited all zones. Students accessed context-aware contents related to the location and engaged in context-aware learning. Finally, a survey for students was administered and subsequent interviews undertaken upon course completion.

5 Results and Discussion

Following completion of the experiment, a questionnaire survey was given to twenty students to understand their opinions. A seven-point Likert-scale was used for all questions: 1 denotes strong disagreement, while 7 denotes strong agreement, respectively. A total of twenty valid questionnaires were returned, a response rate of 100%. Additionally, statistical analysis was performed to determine the degree of perceived usefulness, ease of use, and attitudes toward the use of the HELLO system.

The responses to item A1 indicated that most students believed that the HELLO system is easily used (m = 6.10). Responses to item A2 (m = 5.80) indicated that the system functions were convenient and sufficient for learning. Students also commented that the tools had many functions with user-friendly interfaces that could assist them in completing the targeted learning activities.

Response results of item B1 (m = 6.05) indicated that the HELLO system can increase the motivation to learn. Many students commented that they were highly motivated to use modern devices such as PDAs. One student stated, "I used the m-Player to learn audio materials for actual circumstances. These interesting experiences could not possibly be learned in textbooks." Results of items B2, B3, and B4 (m = 6.20, 5.40, and 6.45, respectively) indicated that the HELLO system can enhance listening, speaking, and reading skills. Many students viewed the learning activities as engaging and interesting. One student said, "I could practice listening and speaking for actual circumstances. That was an interesting experience." Results of item B5 (m = 2.65) indicated that the HELLO system cannot improve writing skills. One student stated "Writing English on a keyboard-less PDA is inconvenient." Therefore, how to improve the writing skills of students is a major challenge of future studies.

The responses to item C1 (m = 6.20) indicated that most students liked using the HELLO system to learn after class. One student stated, "I could not only read web page-based English materials in campus, but could also listen to audio materials anywhere." Responses to item C2 (m = 6.15) indicated that most students would like to use the HELLO system in other courses.

Responses to item D1 (m = 5.90) indicated that the VLT appears to be part of the real world. Responses to items D2 and D3 (m = 6.10 and 6.15, respectively) indicated that the VLT is helpful and can enhance learning experiences. Many students commented that they enjoyed watching a virtual tutor on their PDAs. One student stated, "Through the m-Speaker function, I could see the virtual tutor and listen to English materials, which was a novel experience."

This study evaluated the internal consistency reliability of the questionnaire by using Cronbach's alpha coefficient. Table 4 reveals that all Cronbach's alphas of group A (0.76), B (0.75), C (0.76), and D (0.73) in this experiment exceeded 0.7, indicating the high reliability of the administered questionnaire.

Table 4. Summary of survey results from twenty students (7-point Likert scale)

Group	Item	Mean	SD
A. Easiness	A1. The user interface of u-Tools is friendly.	6.10	0.72
	A2. The functions of the m-Tools are sufficient.	5.80	0.77
B. Usefulness	B1. Applying the task-based HELLO to assist English learning can increase my learning interest and motivation.	6.05	0.76
	B2. Applying the HELLO to assist English learning can increase my listening ability.	6.20	0.70
	B3. Applying the HELLO to assist English learning can increase my speaking ability.	5.40	0.68
	B4. Applying the HELLO to assist English learning can increase my reading ability.	6.45	0.60
	B5. Applying the HELLO to assist English learning can increase my writing ability.	2.65	0.93
C. Attitude	C1. I like to use the HELLO to assist English learning after class.	6.20	0.61
	C2. I hope other courses can also use the HELLO to assist learning.	6.15	0.75
D. Usefulness of VLT	D1. The VLP seems to be part of the real world.	5.90	0.64
	D2. The VLP is helpful for completing the learning activity.	6.10	0.79
	D3. The VLP is helpful for learning.	6.15	0.67

d.f. (degree of freedom) = 19.

To include virtual images and actual scenes on the PDA screens, we recommend using a proper 2D barcode tag size to increase the effectiveness of the HELLO system. When a smaller 2D barcode tag is used, the distance between the tag and phone must be shortened, subsequently making the scene small and unclear. In our experience, proper length of a 2D barcode ranges between 8 and 12 cm, while the proper distance between the tag and phone is 40 to 50 cm. The appropriate angle between normal vector of wall and camera ranges from 0 to 30 degrees. With respect to system constraints, the PDA phone has a small screen size and unclear display under strong sunlight in an outdoor environment.

6 Conclusions

This work has developed a 2D barcode, handheld, augmented reality (AR)-supported English learning environment, called handheld English language learning organization

(HELLO), which provides valuable learning resources and functions to facilitate English language learning. HELLO consists of two subsystems: the HELLO server, and m-Tools. Teachers connect with the HELLO server using their personal computers via the Internet. Students communicate with the HELLO server using their mobile phones via WLAN. A pilot study was performed with the participation of three instructors and twenty undergraduate students from the National Taipei University of Technology (NTUT). Learning activities were conducted in the university. Mobile context-aware and task-based learning pedagogical strategies were adopted. An independent study activity entitled "My Student Life" and a context-aware learning activity called "Campus Tour" were undertaken in a four-week course. "My Student Life" is a learning game. The campus map had many zones marked on the map. The students simply clicked the desired zone and, then, automatically opened materials related to that zone. "Campus Tour" is an augmented reality game. When approaching a zone, a student used the PDA phone to decrypt the 2D barcode and then obtained context-aware contents from server. The students then practiced conversation with the virtual learning partners (tutors).

A questionnaire was administered to the students upon conclusion of learning activities. Based on those results, most students found the course interesting. Responses indicate that most students found HELLO easy to use and useful for assisting learning; they thus endorsed the use of HELLO in future learning. Analysis results also indicate that HELLO not only increased students' motivation to learn, but also enhanced their learning outcomes. This study further demonstrates that 2D barcodes and handheld AR technologies are useful in providing context-aware, immersive experiences in English-learning activities.

Future studies should strive to enhance new sensor networks, physical interaction, and ubiquitous AR technologies. We will continuously work with university English instructors to conduct full-scale studies and to investigate the feasibility of HELLO in various campus contexts and adapt HELLO to individual students' needs, interests, styles, and learning capacity.

References

1. Ogata, H., Yano, Y.: Knowledge Awareness for a Computer-assisted Language Learning using Handhelds. International Journal of Continuous Engineering Education and Lifelong Learning 14, 435–449 (2004)
2. Uther, M., Zipitria, I., Singh, P., Uther, J.: Mobile Adaptive CALL (MAC): A Case-study in Developing a Mobile Learning Application for Speech/Audio Language Training. In: Proceedings of the 3rd International Workshop on Wireless and Mobile Technologies in Education (WMTE 2005), pp. 187–191 (2005)
3. Tan, T.H., Liu, T.Y.: The MObile-Based Interactive Learning Environment (MOBILE) and a Case Study for Assisting Elementary School English Learning. In: Proceedings of 2004 IEEE International Conference on Advanced Learning Technologies (ICALT 2004), Joensuu, Finland, pp. 530–534 (2004)

4. Cooper, J.: Engaging the [Museum] visitor: Relevance, Participation and Motivation in Hypermedia Design. In: Proceedings of 2nd International Conference on Hypermedia and Interactivity in Museums, pp. 174–177 (1993)
5. Jones, V., Jo, J.H.: Ubiquitous Learning Environment: An Adaptive Teaching System Using Ubiquitous Technology. In: Proceedings of the 21st ASCILITE (Australasian Society for Computers in Learning in Tertiary Education) Conference, pp. 468–474. ASCILITE, Perth (2004)
6. Schilit, B., Adams, N., Want, R.: Context-aware Computing Applications. In: IEEE Workshop on Mobile Computing Systems and Applications (WMCSA 1994), Santa Cruz, CA, US (1994)
7. Dey, A.K.: Understanding and Using Context. Personal and Ubiquitous Computing 5(1), 4–7 (2001)
8. Ogata, H., Akamatsu, R., Yano, Y.: TANGO: Computer Supported Vocabulary Learning with RFID tags. Journal of Japanese Society for Information and Systems in Education 22(1), 30–35 (2005)
9. Billinghurst, M., Kato, H., Poupyrev, I.: The MagicBook: A Transitional AR Interface. Computers & Graphics 25(5), 745–753 (2001)
10. Li, L., Zheng, Y., Ogata, H., Yano, Y.: A Framework of Ubiquitous Learning Environment. In: Proceeding of the 4th International Conference on Computer and Information Technology (CIT 2004), Wuhan, China (2004)
11. Li, L., Zheng, Y., Ogata, H., Yano, Y.: A Conceptual Framework of Computer-supported Ubiquitous Learning Environment. Journal of Advanced Technology for Learning 2(4), 187–197 (2005)
12. Liu, T.Y., Tan, T.H., Chu, Y.L.: 2D Barcode and Augmented Reality Supported English Learning System. In: Proceedings of the 6th IEEE International Conference on Computer and Information Science (ICIS 2007), Melbourne, Australia, pp. 5–10 (2007)
13. Hughes, C.E., Stapleton, C.B., Hughes, D.E., Smith, E.M.: Mixed Reality in Education Entertainment, and Training. IEEE Computer Graphics and Applications 26(6), 24–30 (2005)
14. University of Washington, http://www.hitl.washington.edu/artoolkit
15. ARToolworks, Inc., http://www.artoolworks.com/Products.html
16. Graz University of Technology, http://studierstube.icg.tu-graz.ac.at/handheld_ar/
17. ARToolKitPlus, http://studierstube.icg.tu-graz.ac.at/handheld_ar/artoolkitplus.php
18. Studierstube, http://studierstube.icg.tu-graz.ac.at/handheld_ar/stbtracker.php
19. Davis, F.D.: A technology Acceptance Model for Empirically Testing New End-user Information Systems: Theory and Results, Doctoral Dissertation, Sloan School of Management, Massachusetts Institute of Technology, Cambridge, MA (1986)
20. Davis, F.D., Bagozzi, R.P., Warshaw, P.R.: User Acceptance of Computer Technology: A Comparison of two Theoretical Models. Management Science 35, 982–1003 (1989)
21. Davis, F.D.: User Acceptance of Information Technology: System Characteristics, User Perceptions and Behavioral Impacts. International Journal Man-Machine Studies 38, 475–487 (1993)
22. Adams, D.A., Nelson, R.R., Todd, P.A.: Perceived Usefulness, Ease of Use, and Usage of Information Technology: A replication. MIS Quarterly 16(2), 227–247 (1992)

23. Hendrickson, A.R., Massey, P.D., Cronan, T.P.: On the Test-retest Reliability of Perceived Usefulness and Perceived Ease of Use Scales. Management Information Systems Quarterly 17, 227–230 (1993)
24. Szajna, B.: Software Evaluation and Choice: Predictive Evaluation of the Technology Acceptance Instrument. Management Information Systems Quarterly 18(3), 319–324 (1994)
25. Nunnaly, J.: Psychometric Theory. McGraw-Hill, New York (1978)
26. Ellis, R.: Task-based Language Learning and Teaching. Oxford University Press, New York (2003)
27. Johnson, D.W., Johnson, R.T.: Learning Together and Alone: Cooperative, Competitive, and Individualistic Learning. Allyn and Bacon, Boston (1994)
28. Nunan, D.: Designing Tasks for the Communicative Classroom. Cambridge University Press, London (1989)
29. Nunan, D.: Research Methods in Language Learning. Cambridge University Press, Cambridge (1992)
30. Willis, J.: A Framework for Task-based Learning. Longman, London (1996)
31. Carless, D.: Issues in Teachers' Reinterpretation of a Task-based Innovation in Primary Schools. TESOL Quarterly 38(4), 639–662 (2004)

Robust 1-D Barcode Recognition on Camera Phones and Mobile Product Information Display

Steffen Wachenfeld, Sebastian Terlunen, and Xiaoyi Jiang

Department of Computer Science, University of Münster, Germany
{wachi,terlunen,xjiang}@uni-muenster.de

Abstract. In this paper we present a robust algorithm for the recognition of 1-D barcodes using camera phones. The recognition algorithm is highly robust regarding the typical image distortions and was tested on a database of barcode images, which covers typical distortions, such as inhomogeneous illumination, reflections, or blurs due to camera movement. We present results from experiments with over 1,000 images from this database using a MATLAB implementation of our algorithm, as well as experiments on the go, where a Symbian C++ implementation running on a camera phone is used to recognize barcodes in daily life situations. The proposed algorithm shows a close to 100% accuracy in real life situations and yields a very good resolution dependent performance on our database, ranging from 90.5% (640 × 480) up to 99.2% (2592 × 1944). The database is freely available for other researchers. Further we shortly present MobilePID, an application for mobile product information display on web-enabled camera phones. MobilePID uses product information services on the internet or locally stored on-device data.

1 Introduction

Barcodes are ubiquitously used to identify products, goods or deliveries. Reading devices for barcodes are all-around, in the form of pen type readers, laser scanners, or LED scanners. Camera-based readers, as a new kind of barcode reader, have recently gained much attention. The interest in camera-based barcode recognition is build on the fact that numerous mobile devices are already in use, which provide the capability to take images of a fair quality. In combination with Bluetooth or WLAN connectivity, new applications become possible, e.g. an instant barcode-based identification of products and the online retrieval of product information. Such applications allow for the display of warnings for people with allergies, results of product tests or price comparisons in shopping situations.

We have created a database of barcode images taken by a cell-phone camera using different resolutions. This database was used to create a robust algorithm using MATLAB. The algorithm was then ported to Symbian C++ to allow for a fast execution on Symbian OS powered devices and access to the devices' GPS, Bluetooth and WLAN functionalities. We mainly used a Nokia N95 camera

X. Jiang, M.Y. Ma, and C.W. Chen (Eds.): WMMP 2008, LNCS 5960, pp. 53–69, 2010.
© Springer-Verlag Berlin Heidelberg 2010

Fig. 1. The Nokia N95 camera phone which supports both Java and Symbian OS applications

phone (see Figure 1) which supports both Java and Symbian OS and has GPS and WLAN capabilities. This allows to compare the recognition performance with existing Java or C++ barcode recognition tools and to create powerful applications, which for example retrieve product information using a WLAN internet connection.

Efforts concerning the recognition of 1-D barcodes using camera phones have already been made. Adelmann et al. [1] have presented two prototypical applications: the display of literature information about scanned books, and the display of ingredient information about scanned food for allergic persons. They did not report recognition performances, but showed proof of concept for new applications. Wang et al. from Nokia [11,12] have presented an algorithm which seems very fast but also very simple. They reported a recognition rate of 85,6% on an unpublished image database. From their description we consider the algorithm to be much less robust than the algorithm proposed in this paper. Further, Ohbuchi et al. [6] have presented a real-time recognizer for mobile phones, which is assembler-based and kept very simple. Chai and Hock [3] have also presented an algorithm, but without specifying recognition results. Early barcode recognition algorithms (e.g. Muniz et al. [5], and Jospeh and Pavlidis [4]) and parts of the mentioned approaches achieve their goal by applying techniques like Hough transformation, wavelet-based barcode localization, or morphological operations. Using such techniques leads to computationally expensive implementations, which may be not well suited for the use with mobile devices.

In this paper, we firstly present an algorithm to recognize 1-D barcodes, which works for the widely used standards UPC-A, EAN-13 and ISBN-13. Our algorithm uses image analysis and pattern recognition methods which rely on knowledge about structure and appearance of 1-D barcodes. Given the computational power and the image quality of today's camera phones, our contribution is an algorithm which is both fast and robust. Secondly, we present our software's capability to use the recognized barcode to look up and display product information from internet data services or on-device data.

We start with a brief description of barcodes in Section 2. The details of our algorithm are explained in Section 3. In Section 4 we present results of experiments using our database and of experiments on the go, where the Symbian C++ implementation running on a camera phone is used to recognize barcodes in daily life situations. In Section 5 we shortly explain how our software uses internet services or on-device data to display information on scanned products. Finally, Section 6 gives a conclusion and outlines future work.

2 Brief Description of Barcodes

All barcode recognition algorithms explore to some extent the knowledge about the UPC-A/EAN-13/ISBN-13 barcodes. In this section we give a brief description.

These barcodes consists of 13 digits and their last digit is a checksum which is computed from the first 12 digits. The barcode starts by a left-hand guard bar A (black-white-black) and ends with a right-hand guard bar E (black-white-black) (see Figure 2). Between the guard bars, there are two blocks B and D of 6 encoded digits each, separated by a center bar C (white-black-white-black-white).

The smallest unit is called a module. Bars and spaces can cover one to four modules of the same color. Each digit is encoded using seven modules (two bars and two spaces with a total width of 7 modules). The width of a complete UPC-A/EAN-13/ISBN-13 barcode is 59 black and white areas $(3 + 6*4 + 5 + 6*4 + 3)$ which consist of 95 modules $(3 + 6*7 + 5 + 6*7 + 3)$.

Two alphabets can be used to encode a digit, the even alphabet or the odd alphabet. While the last 12 digits are directly coded using these two alphabets,

Fig. 2. Structure of UPC-A/EAN-13/ISBN-13 barcodes: Left-/right guard bars (A/E), center bar (C), blocks of the encoded digits (B/D), meta-number, and checksum

Fig. 3. EAN-13/ISBN-13 alphabets and meta-number calculation

the first of the 13 digits is determined by the alphabets that have been used to encode the first six digits. Thus, the first digit is called meta-number or induced digit (see Figure 3).

3   Recognition Algorithm

The goal of our algorithm is to be both fast and robust. Although we have implemented and experimented with features such as automatic rotation and perspective transformations, we do not make use of them here for performance reasons. Instead, we expect an image containing a 1-D barcode which covers the image center as input. The barcode does not need to be centered, may be upside down, or may have the usual perspective distortions and rotations (approx. ±15 degrees) that occur when images are taken by camera phones.

In Figure 4 the three phases of the developed recognition algorithm and the corresponding steps are shown. The algorithm can be devided in three major phases:

– preprocessing and binarization
– digit classification
– search for the most similar code.

These phases are explained in the following.

3.1   Phase I: Preprocessing and Binarization

Other approaches start with global smoothing, wavelet-based barcode area lo-cation [11] or even morphological operations [3]. We consider this as being too time-consuming and developed a fast but robust scanline-based approach.

We assume that a horizontal scanline in the middle of the image covers the barcode. If this is not the case, or if parts of the barcode which lay on the scanline are dirty, occluded or affected by strong reflections, we will detect this in a very early stage and will repeat our algorithm using alternate scanlines above and below.

The first step is thus to extract the scanline which is a single line of pixels. This is the centerline in the first attempt or may be a line above or below in a subsequent attempt.

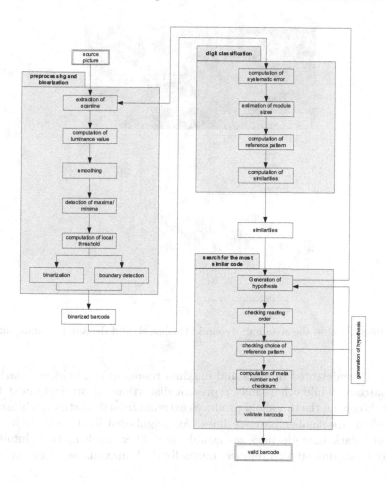

Fig. 4. The three phases of the developed recognition algorithm and the corresponding steps

For the binarization a dynamic threshold is required, to be robust against dirt, badly printed barcodes or illumination changes. To do this we compute the luminance value $Y(x) \in [0..1]$ for each position x on the scanline ($Y(x) = 0.299R(x) + 0.587G(x) + 0.114B(x)$).

In the next step the luminance values on the scanline are smoothed using a small Gaussian kernel. Instead of the whole image, just the pixels on the scanline are smoothed.

Our dynamic thresholding algorithm for binarization depends on maxima and minima which are searched in the next step. We search for local minima and maxima along the scanline, so that neighbored minima and maxima have a luminance difference $\Delta Y \geq 0.01$.

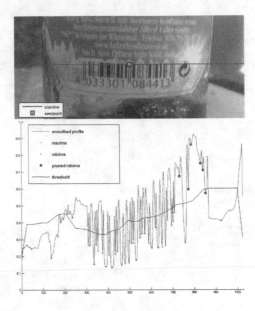

Fig. 5. Intensities on the scanline, dynamic threshold, and detected minima/maxima

Some of the detected minima and maxima represent the light and dark lines of the barcode, while others may represent distortions or artefacts next to the barcode. Lowering the threshold produces extrema from distortions, while raising it may eliminate smaller barcode lines. As neighbored light barcode lines and neighbored dark barcode lines do usually not differ much in their luminance, we apply a pruning step to remove unusually dark maxima and unusually light minima.

After the pruning step, the next and biggest step is the computation of the local thresholds. For each position x from the middle of the scanline to the image border, the threshold $t(x)$ for binarization is computed by evaluating a function depending on the last inward seven minima/maxima. We do not consider the outer minima/maxima, as we do not want regions outside the barcode to impact thresholds within the barcode area. Figure 5 shows the profile, minima and maxima, pruned extrema and the resulting threshold. The evaluated function averages the luminance value of the second lowest maxima and of the second highest minima to achieve a good robustness against local errors like dirt or heavy noise.

Investigating a number of seven minima/maxima means to always consider the bars/spaces of the current digit and three other digits. This prevents small bars/spaces from becoming victim of this implicit outlier detection. This step and some of the following steps make use of knowledge about the UPC-A/EAN-13/ISBN-13 barcodes (see Section 2).

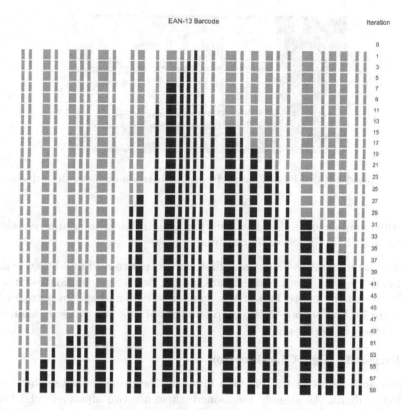

Fig. 6. Barcode boundary detection which starts at the center and successively adds white-black area pairs and looks for guard bars in order to terminate if the number of areas between two guard bar candidates is 59

The last step is a combined binarization and barcode boundary detection and makes use of the facts described in Section 2. The barcode boundary detection starts at a seed point in the middle of the scanline. We want the starting position to be a bar-pixel, so either the seed point is black $(Y(x) < t(x))$ or we start with the closest point to the left or right which is black. The combined binarization and boundary detection algorithm considers pairs of white-black areas and works as follows (see Figure 6):

1. Binarization of the next white-black area pairs to the left and to the right using the dynamic threshold and remembering their sizes.
2. Adding the white-black area pair of that side, which has the smaller white area and whose black area is not larger than 8 times the smallest black area. (If the black areas on both sides are too large: FAIL and try different seed point or alternate profile, as this is considered as an exceed of the area size maximum).
3. Marking the added area pair as part of a possible guard bar if the last three black/white/black areas are approximately single modules.

Fig. 7. Guard bar candidates found during the barcode bounday detection step

4. If the number of areas between two guard bar candidates equals 59, terminate: SUCCESS - else: continue from step 1.

By adding smaller white-black pairs first, we prevent to add non-barcode areas. By remembering the sizes of added bars, we can determine candidates for guard bars. We found the guard bars if the number of bars and spaces between is 59 (see Figure 7). The output of this step and thus of this first phase is a binarized barcode image of height one.

3.2   Phase II: Digit Classification

The bars and spaces of the barcode found in the previous phase, have to be classified as digits. As already mentioned, there are two alphabets of 10 digits each, which results in 20 classes c_i. Each digit s is encoded using two black and white areas having a total width of 7 modules.

As preparation for the classification, we want to generate an image specific prototype for each class which reflects a systematic error due to illumination. First we need to determine the width of a single module. To do this, we investigate the two guard bars on the left and right as well as the center bar. Since we know that their black and white areas have the width of one module each, we can compute the average widths of single black and white modules. From the width of single modules we can compute the width of the double, triple and quadruple modules. Please note that in contrast to what could be expected, the binarized area of a double module is not twice as large as the area of a single module. Further, the width of a single black module is not the barcode's total width divided by 95. The average (single-) module width in pixels $\overline{m_1}$ can be approximated as the mean of all module widths $|m_i|$. It can be derived directly from the width of whole barcode $|\vec{c}|$.

$$\overline{m_1} = \frac{\sum\limits_{i=1}^{95} |m_i|}{95} = \frac{|\vec{c}|}{95} \tag{1}$$

Due to brighter or darker illumination the average widths of single white $\overline{w_1}$ and black $\overline{b_1}$ modules will differ from $\overline{m_1}$. The average widths of single white modules

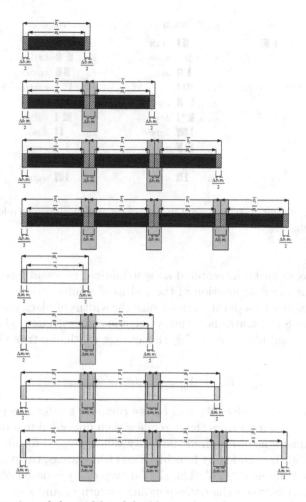

Fig. 8. Computation of the widths of double, triple and quadruple modules, here assuming a darker image where $\overline{w_1} < \overline{m_1} < \overline{b_1}$

$\overline{w_1}$ and of a single black modules $\overline{b_1}$ can be estimated using the modules of the guard and center bars. All areas a_i of these bars are single modules, so that

$$\overline{b_1} = \frac{|a_1| + |a_3| + |a_{29}| + |a_{31}| + |a_{57}| + |a_{59}|}{6} \tag{2}$$

$$\overline{w_1} = \frac{|a_2| + |a_{28}| + |a_{30}| + |a_{32}| + |a_{58}|}{5}. \tag{3}$$

This allows to consider the following differences

$$\Delta_{\overline{m_1}\,\overline{w_1}} = \overline{m_1} - \overline{w_1} \tag{4}$$

$$\Delta_{\overline{b_1}\,\overline{m_1}} = \overline{b_1} - \overline{m_1} \tag{5}$$

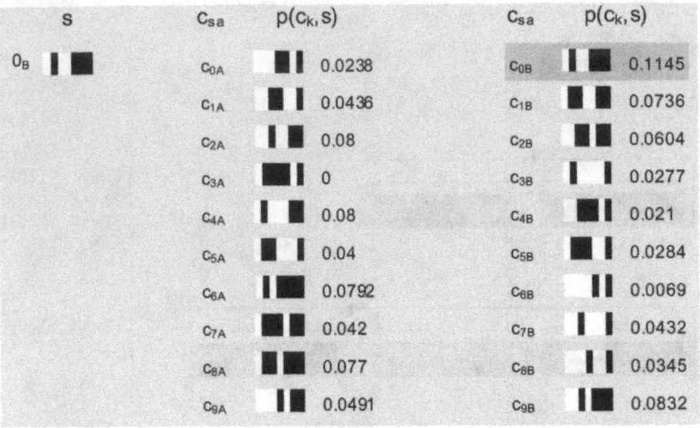

s	C_{sa}	$p(c_k,s)$		C_{sa}	$p(c_k,s)$	
0_B	c_{0A}		0.0238	c_{0B}		0.1145
	c_{1A}		0.0436	c_{1B}		0.0736
	c_{2A}		0.08	c_{2B}		0.0604
	c_{3A}		0	c_{3B}		0.0277
	c_{4A}		0.08	c_{4B}		0.021
	c_{5A}		0.04	c_{5B}		0.0284
	c_{6A}		0.0792	c_{6B}		0.0069
	c_{7A}		0.042	c_{7B}		0.0432
	c_{8A}		0.077	c_{8B}		0.0345
	c_{9A}		0.0491	c_{9B}		0.0832

Fig. 9. Example of digit classification: Query image s and similarity to reference pattern of both alphabets

These differences can be interpreted as systematic errors and have to be taken into account for the computation of the widths of double, triple and quadruple modules and require to separately determine the widths of black and white single modules. Figure 8 illustrates how the systematic error effects the beginning and the end of white and black areas. E.g. the average width of a triple black module $\overline{b_3}$ is

$$\overline{b_2} = \overline{m_3} + \Delta_{\overline{b_1}\ \overline{m_1}} = 3\overline{m_1} + \Delta_{\overline{b_1}\ \overline{m_1}}. \tag{6}$$

Based on the average widths $\overline{w_1}$ and $\overline{b_1}$, we compute a reference pattern r_k for each class c_k. Each digit s from the barcode is then presented to a distance based classifier which assigns normalized similarity values $p(c_k, s)$ for all classes c_k.

Since each digit is encoded by four black and white areas, the pattern r can be represented as 4-tupels $r \in \mathbb{R}^4$. The similarity $p(c_k, s)$ is based on the squared distance $d(r_k, r_s)$ between the corresponding pattern r_k and r_s:

$$p'(c_k, s) = 1 - \frac{d(r_k, r_s)}{\max_i(d(r_i, r_s))} \tag{7}$$

$$p(c_k, s) = \frac{p'(c_k, s)}{\sum_{i=1}^{20} p'(c_i, s)} \tag{8}$$

Figure 9 shows an example of digit classification. The output of this phase are normalized similarity values for each digit to each class.

3.3 Phase III: Search for the Most Similar Code

We now have similarity values for each digit s to each class c_k and our goal is to determine the most likely barcode that is valid. By combining the results of the twelve encoded digits $s^{(1)}, s^{(2)}, \ldots, s^{(12)}$ we can successively generate code hypotheses $(m, c^{(1)}, c^{(2)}, \ldots, c^{(12)})$, where m is the meta number (see Figure 10).

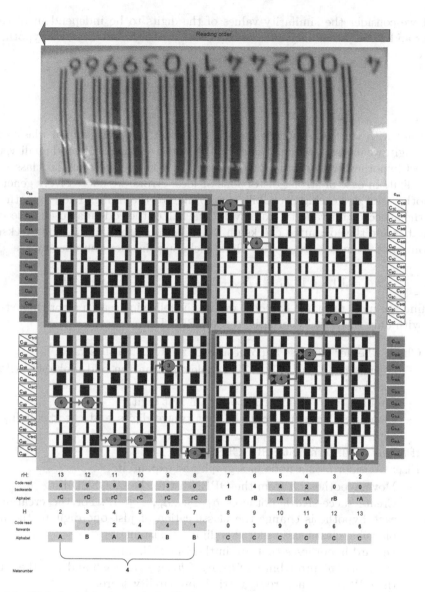

Fig. 10. Code hypothesis as path. Each row represents a class. The most plausible class is marked. The result is a path that can be assigned a total probability. The goal is to find the most probable and valid path.

The result of the third and last phase is thus the decoded barcode and a corresponding probability value which represents the certainty.

If we consider the similarity values of the digits to be independent of each other and to be probability like, we can consider the probability of a hypothesis to be:

$$p(c^{(1)}, \ldots, c^{(12)} | s^{(1)}, \ldots, s^{(12)}) = \prod_{i=1}^{12} p(c^{(i)} | s^{(i)})^{1/12} \qquad (9)$$

Starting with the code hypothesis resulting from the most similar classes for each digit, we investigate hypotheses on-demand. For images of good quality, the correct hypothesis is normally the one resulting from the most similar classes for each digit. In case of strong distortions or local errors, we successively generate hypotheses of decreasing probability, according to the input. These hypotheses are then checked for validity. To be valid one half of the code may use the second alphabet and the computed checksum has to be equal to digit 13. The checksum is computed from the first twelve digits including the meta-number by

$$s_{13} = 10 - ((1 \cdot s_1 + 3 \cdot s_2 + 1 \cdot s_3 + \ldots + 3 \cdot s_{12}) \bmod 10). \qquad (10)$$

To find the most probable barcode, we use the following search algorithm, starting with only the best hypothesis in the OPEN-List:

1. Check whether the first hypothesis h in the OPEN-list is valid:
   - Is the second half (D) of h completely from the odd alphabet? (if no: read backwards)
   - Compute meta-number m to obtain first digit.
   - Compute checksum from the first twelve digits and check equality to digit 13.
2. If h was valid, terminate: SUCCESS.
   Else: Consider h as a node in a graph and expand it:
   - Move hypothesis h from the OPEN-list to the CLOSED-list.
   - Create successive hypotheses $h_1, h_2, \ldots h_n, n \leq 12$: For the creation of each hypothesis change the classification of just one digit to the next plausible classification for this digit (if existent).
   - Discard hypotheses that are in the CLOSED-List.
   - Compute the probability of $h_1, h_2, \ldots h_n, n \leq 12$, each and sort them into the OPEN-List according to their probability score.
3. Continue from step 1.

4 Experimental Results

Here, we distinguish between two kinds of experiments. First, experiments based on a database, where a MATLAB based algorithm is applied to a fixed number of images, taken by a camera phone. And second, experiments on the go, where a Symbian C++ implementation running on a camera phone is used to recognize barcodes in daily life situations.

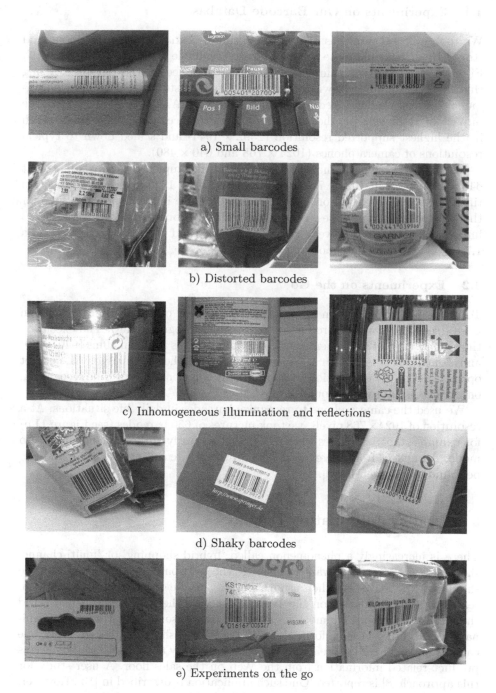

Fig. 11. a)-d): Examples of pictures stored in the database. e) Examples from the experiments on the go.

4.1   Experiments on Our Barcode Database

We took over 1,000 images of barcodes using a Nokia N95 camera phone and stored them in a database which is freely available to other researchers [9]. Figures 11 a)-d) give examples of pictures contained in the database. Using this fixed set of images, which includes heavily distorted, shaky and out-of-focus pictures, we developed our algorithm using MATLAB. To investigate the impact of resolution on the recognition performance, the images were taken in the highest supported resolution (2592 × 1944) and scaled down to typical resolutions of camera phones (1024 × 768 and 640 × 480).

The recognition performance of our MATLAB implementation is 90.5% at 640 × 480, 93.7% at 1024 × 768, and 99.2% at 2592 × 1944 pixels. The execution time varies due to resolution, image quality, and the image dependent time for the generation of hypotheses. For the recognition of one barcode, our MATLAB implementation takes between 25ms (640 × 480) and 50ms (2592 × 1944) on a 1,7 GHz single-core notebook.

4.2   Experiments on the Go

To perform experiments on the go, we implemented our algorithm in Symbian C++ (see [7,10] for an introduction to Symbian programming). The execution time of our current implementation for Symbian OS is between 50ms (640 × 480) and 70ms (1024 × 768) on a Nokia N95 camera phone. Our code is not yet optimized, but the recognition speed is already sufficient for real-time recognition on a video stream with about 20fps.

We used the camera phone to recognize barcodes in daily life situations. At a resolution of 1024 × 768 pixels, we took pictures of 650 barcodes (see Figure 11 e) for examples). 650 barcodes were recognized correctly without using the macro mode. All images from the experiments on the go are collected, stored, and can be found on our website [9].

5   Product Information Display

There is increasingly a phenomenon called 'hybrid shopping' or 'multi-channel shopping'. As consumers want to leverage the transparency in Internet and the shopping experience from their usual shopping, people go back and forth between searching for information on the Web and buying their products in real world shops. Modern mobile technologies enable a time-wise and location-wise combination of Internet-based information and traditional shopping. For example, one can use barcode reading on a mobile phone to identify a product and then access product-related information from the Web on the sales floor. A user study for this approach [8] is reported. One such application is described in [13]. However, the barcode reading is not done on a mobile device, but on a desktop computer. Another related application is the mobile electronic patient diaries reported in [2]. But the barcode reading is realized by an external barcode scanning device.

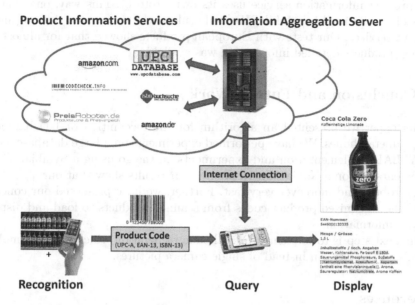

Fig. 12. Mobile recognition and information display for products with 1-D barcodes

In all such applications a robust barcode recognition directly on mobile phones is essential.

We have implemented a distributed application for product information display which consists of a mobile client and an internet server. Figure 12 demonstrates the process of product recognition and information display, which will be explained in the following.

The mobile client is a Symbian OS application named *MobilePID*, which stands for 'Mobile Product Information Display'. MobilePID uses our camera-based barcode recognition algorithm to identify products by their 1-D barcode. Scanned product codes, e.g. UPC-A, EAN-13, ISBN-13, as well as manually entered codes are used by MobilePID to look up and display product information.

If the mobile device is offline, the information contained in the barcode itself (e.g. manufacturer and country of origin) and information from a local on-device database is displayed. The local on-device database is customizable and may contain information on frequent or user specified products, such as books, videos or certain foods.

On web-enabled devices, MobilePID uses the internet connection to download information from our internet server. Our internet server is an information aggregation server, which itself consults several available product information services on the internet. The server aggregates the information it gets from the different services, such as *upcdatabase.com*, *codecheck.info*, *amazon*, *ISBN book search*, and many more. We use the intermediate aggregation server because each product information service uses its own protocol. This way, one general protocol can be used by MobilePID to download the aggregated data of many different services. Our tests with European products showed that for almost all scanned products detailed information was available.

6   Conclusion and Future Work

In this paper we presented an algorithm for the recognition of 1-D barcodes using camera phones. We have performed experiments on a large database using a MATLAB implementation and experiments on the go using a Symbian C++ implementation on a Nokia camera phone. Our results show that our algorithm is very robust and moreover very fast. Further, we have presented our concept to use the recognized product codes from scanned products to load and display product information.

Our next step is to implement a real-time recognition which uses the mobile phone's camera stream instead of single camera pictures.

References

1. Adelmann, R., Langheinrich, M., Flörkemeier, C.: Toolkit for bar code recognition and resolving on camera phones – Jump starting the Internet of things. In: Workshop on Mobile and Embedded Interactive Systems, MEIS 2006 (2006)
2. Arens, A., Rösch, N., Feidert, F., Harpes, P., Herbst, R., Mösges, R.: Mobile electronic patient diaries with barcode based food identification for the treatment of food allergies. GMS Med. Inform. Biom. Epidemiol. 4(3), Doc14 (2008)
3. Chai, D., Hock, F.: Locating and decoding EAN-13 barcodes from images captured by digital cameras. In: 5th Int. Conf. on Information, Communications and Signal Processing, pp. 1595–1599 (2005)
4. Joseph, E., Pavlidis, T.: Bar code waveform recognition using peak locations. IEEE Trans. on PAMI 16(6), 630–640 (1998)
5. Muniz, R., Junco, L., Otero, A.: A robust software barcode reader using the Hough transform. In: Proc. of the Int. Conf. on Information Intelligence and Systems, pp. 313–319 (1999)
6. Ohbuchi, E., Hanaizumi, H., Hock, L.A.: Barcode readers using the camera device in mobile phones. In: Proc. of the Int. Conf. on Cyberworlds, pp. 260–265 (2004)
7. Rome, A., Wilcox, M.: Multimedia on Symbian OS: Inside the Convergence Device. John Wiley & Sons, Chichester (2008)
8. von Reischach, F., Michahelles, F., Guinard, D., Adelmann, R., Fleisch, E., Schmidt, A.: An evaluation of product identification techniques for mobile phones. In: Gross, T., Gulliksen, J., Kotzé, P., Oestreicher, L., Palanque, P., Oliveira Prates, R., Winckler, M. (eds.) INTERACT 2009. LNCS, vol. 5727, pp. 804–816. Springer, Heidelberg (2009)
9. Wachenfeld, S., Terlunen, S., Madeja, M., Jiang, X.:
   http://cvpr.uni-muenster.de/research/barcode
10. Wachenfeld, S., Madeja, M., Jiang, X.: Developing mobile multimedia applications on Symbian OS devices. In: Jiang, X., Ma, M.Y., Chen, C.W. (eds.) WMMP 2008. LNCS, vol. 5960, pp. 238–263. Springer, Heidelberg (2010)

11. Wang, K., Zou, Y., Wang, H.: Bar code reading from images captured by camera phones. In: 2nd Int. Conf. on Mobile Technology, Applications and Systems, pp. 6–12 (2005)
12. Wang, K., Zou, Y., Wang, H.: 1D bar code reading on camera phones. Int. Journal of Image and Graphics 7(3), 529–550 (2007)
13. Wojciechowski, A., Siek, K.: Barcode scanning from mobile-phone camera photos delivered via MMS: Case study. In: Song, I.-Y., Piattini, M., Chen, Y.-P.P., Hartmann, S., Grandi, F., Trujillo, J., Opdahl, A.L., Ferri, F., Grifoni, P., Caschera, M.C., Rolland, C., Woo, C., Salinesi, C., Zimányi, E., Claramunt, C., Frasincar, F., Houben, G.-J., Thiran, P. (eds.) ER Workshops 2008. LNCS, vol. 5232, pp. 218–227. Springer, Heidelberg (2008)

Evaluating the Adaptation of Multimedia Services Using a Constraints-Based Approach

José M. Oliveira[1] and Eurico M. Carrapatoso[2]

[1] INESC Porto, Faculdade de Economia, Universidade do Porto,
[2] INESC Porto, Faculdade de Engenharia, Universidade do Porto,
rua Dr. Roberto Frias, 378, 4200-465 Porto, Portugal
{jmo,emc}@inescporto.pt

Abstract. This chapter presents a proposal to solve the problem of the adaptation of multimedia services in mobile contexts. The chapter combines context-awareness techniques with user interface modeling to dynamically adapt telecommunications services to user resources, in terms of terminal and network conditions. The solution is mainly characterized by the approach used for resolving the existing dependencies among user interface variables, which is based on the constraints theory. The experiments and tests carried out with these techniques demonstrate a general improvement of the adaptation of multimedia services in mobile environments, in comparison to systems that do not dynamically integrate the user context information in the adaptation process.

Keywords: Multimedia adaptation, Dynamic generation, Mobile environments, Context gathering, SMIL.

1 Introduction

The acceptance of new services in the present context of telecommunications will only be effective if the user has the possibility to access them anywhere, in any technological circumstances, even in roaming scenarios. This user requirement places multimedia service providers under the significant challenge of being able to transform their services in order to adapt them to a great variety of delivery contexts. This need for multimedia service adaptation in mobile environments constituted the main motivation for the research work presented in this chapter.

The chapter defines a generic adaptation methodology targeted to adapt multimedia services provided in the context of a telecommunications operator. The proposed methodology follows the approach of dynamically integrating the user context information, which is by nature very changeable in mobile environments, in order to achieve consistent results in the adaptation process. The chapter is organized as follows. Section 2 compares our approach for the dynamic adaptation of multimedia services with some related work. Section 3 presents an adaptation methodology suitable to adapt telecommunications services to different access mechanisms, connectivity capabilities and user preferences.

X. Jiang, M.Y. Ma, and C.W. Chen (Eds.): WMMP 2008, LNCS 5960, pp. 70–88, 2010.
© Springer-Verlag Berlin Heidelberg 2010

Section 4 presents the use of the constraints theory by a *Media Adapter* targeted to the Synchronized Multimedia Integration Language (SMIL). Section 5 evaluates the proposed adaptation methodology and draws some conclusions regarding the qualitative and quantitative aspects of the multimedia adapter, using a case study service. Section 6 reports the chapter main conclusions.

2   Related Work

The problem of adapting multimedia services and presentations has received significant attention from the research community in the last years. *Cuypers* [1] is a research prototype system for the generation of Web-based multimedia presentations following a constraint-based approach, which was the main inspiration for the constraints generation process of our adaptation methodology. A drawback of *Cuypers* is its focus on the delivery of SMIL content for desktop PCs, not exploring other final presentation formats and devices.

In [2] and [3], the problem of delivering multimedia content to different end device types, using a wide range of multimedia formats, is covered. However, the process used by these systems for acquiring the user context is not dynamic. Besides that, these two adaptation systems do not consider the possibility of media adaptation or transformation, adopting the approach of previously defining alternative media, similar to the SMIL `switch` tag.

The SmartRoutaari [4] is another example of a context-sensitive system. SmartRotuaari is operational at the city center of Oulu, in Northern Finland, and comprises a wireless multi-access network (WLAN, GPRS and EDGE), a middleware architecture for service provisioning, a Web portal with content provider interface and a collection of functional context-aware mobile multimedia services. The contextual information gathered by the SmartRoutaari system include time, location, weather, user preferences and presence status. In contrast with our approach, the SmartRoutaari solution for gathering user context information is proprietary and has no facilities for the seamless expansion with independent third party services. This approach is not in line with the current standardization activities on telecommunications service provision, namely the 3GPP IMS standardization effort [5], which promotes the use of open APIs, such as Parlay/OSA, for opening operator networks to trusted third parties.

The use of an abstract multimedia model as the basis for the adaptation process has been a usual approach in this area. Examples of such models are UIML [6], ZyX [7] and DISL [8], which is an extension of a subset of UIML to enhance the support for interaction modeling. A common characteristic of these models is their ambition to be as generic as possible. UIML is even a standardized language, defined within the OASIS consortium. In contrast with those models, the definition of the *MModel* did not have the ambition of being a standardized way to define user interfaces but only had the objective of defining a model simple enough to demonstrate the adaptation methodology.

A particular characteristic of the *MModel* has to do with its target services. The *MModel* was designed to define user interfaces of telecommunications services, provided in the context of telecommunications operators. In addition, the

MModel also gives support to multimedia features, an important requirement for current telecommunications services.

The use of SMIL in the telecommunications field increased significantly with the 3GPP adoption of the 3GPP SMIL Language Profile as the media synchronization and presentation format for the Multimedia Message Service (MMS) [9]. However, the use of SMIL in the context of interactive services is still very limited. In [10], the authors propose several extensions to SMIL to cover input and output capabilities not currently available, such as location-based information, telephony, forms, scripting and tactile feedback. From these capabilities, the support for forms was crucial during the development of the work presented in this chapter. To add forms support to SMIL, we followed the World Wide Web Consortium (W3C) approach of integrating XForms with SMIL through the definition of a joint mapping from *MModel* to these two languages.

Our work is also related to the recent advances in the area of accessing multimedia services in mobile contexts, through small devices [11], [12], [13]. In these works, the adaptation is usually done directly in the content, by adapting the media codification or by transcoding the media elements. Although the methodology we propose is not optimized for small devices, being the main focus on the capability to cover a large range of access devices and also to be open and scalable to support new future device types, the integration of those adaptation techniques in the *SMIL Media Adapter* had a significant impact on the results obtained (see Section 5).

3   A Methodology for the Adaptation of Telecommunications Services

The adaptation methodology presented in this section is based on the *MModel*, an abstract construction that holds an XML description of the user interface [14]. A service is modeled as a set of *panels*, corresponding each one to the user interface of a specific state of the service logic (see Figure 1). These *panel* elements have the particular role of being the elements over which the adaptation methodology will act, adapting their child elements and presenting them as a user interface view in the user terminal.

The *Service Listener* is responsible for filling the *MModel* object with the user interface specification, while the *Media Adapter* is the entity responsible for performing the adaptation of the *MModel* to a specific format (or media type). With the *MModel* set on the *Adaptation System*, the service adaptation process starts. In each step, the adaptation is carried out over only one of the *MModel* parts. For that, the *Service Listener* sets the focus on the *MModel* part that should be adapted (see Figure 1).

Since in mobile computing the user context information is very dynamic by nature, the capability to dynamically detect this information at the beginning of each adaptation cycle is decisive to achieve acceptable results with the adaptation methodology. The adaptation methodology follows the approach of using the Application Programming Interfaces (APIs) offered by Parlay/OSA [15] to

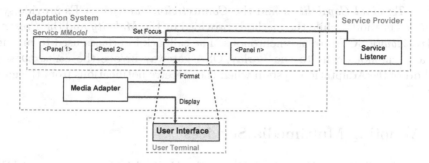

Fig. 1. Structure of an *MModel* representation of a service

obtain user context related information directly from the network. These Parlay/OSA APIs enable telecommunications network operators to open up their networks to third party service providers, in a secure and trusted way, through Parlay/OSA Service Capability Features (SCFs). The SCFs used for obtaining the desired context information were the *Terminal Capabilities* and the *Connectivity Manager* SCFs for the gathering of computing environment information related to the terminal and the network connection used to access the service, and the *Mobility* and the *Presence & Availability* SCFs for the gathering of user environment information related to the location and the status of the user. The process used by the adaptation methodology for gathering the user context is highlighted in Figure 2.

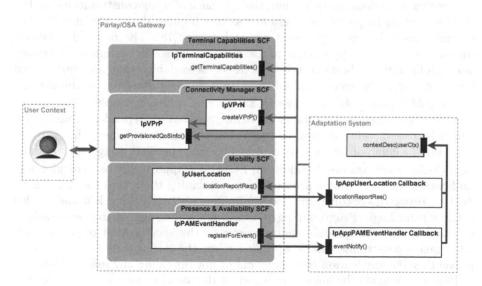

Fig. 2. User context gathering process using Parlay/OSA APIs

The *Terminal Capabilities* and the *Connectivity Manager* SCFs provide synchronous methods, while the methods provided by the *Mobility* and the *Presence & Availability* SCFs are assynchronous. The callback interfaces used to manage the asynchronous calls to the Parlay/OSA gateway invoke the `contextDesc()` method of the Adaptation System when they receive notifications from the gateway.

4   Adapting Multimedia Services

This section discusses the major features of a *Media Adapter* targeted to SMIL, a language defined by W3C, which enables the specification of multimedia presentations for delivery over the Web [16].

The adaptation methodology proposed by us prescribes the definition of the user interface structure, by the service designer, as the initial point of the adaptation process, corresponding to the instantiation of the *MModel* object. Then, the interface structure is maintained stable during the various methodology phases, having no influence on the dynamic change of the user interface during the service session lifetime. External factors, such as the network bandwidth conditions, the time available to see the presentation, the device characteristics where the presentation will be displayed or the user's personal preferences, have a much more active role in the user interface dynamic generation.

The use of constraints by multimedia presentation systems is particularly appropriate to manage the spatial and temporal arrangements of the media items that are part of the interface, since the problem of establishing a final decision concerning the interface spatial and temporal layouts only involves finite and discrete variable domains. Although the approach of using constraints as the basis of a multimedia adaptation system was very popular during the nineties [17], it continues to be used by recent research works [18]. The main idea behind the constraint-based approach is to use a constraint solver system to determine one (preferably the best) solution for a set of variables that are interrelated by constraints. The constraint solver system chosen to be used in the context of the *SMIL Media Adapter* was ECL^{i}PS^e [19], which not only offers a Java interface, but also the possibility to define application-dependent constraints, which is a very useful feature for multimedia applications, where the relations between the different interveners cannot be easily specified using typical numerical domains constraints. In addition, ECL^{i}PS^e supports the backtracking and unification features of logic programming, combining them with the domain reduction properties of constraint programming, resulting in what is usually called a Constraint Logic Programming (CLP) system. The use of the backtracking mechanism gives a dynamic characteristic to the specification of constraints. Alternative constraints can be specified when the initial ones cause a failure, preventing the application from crashing or not providing any information.

Figure 3 presents the algorithm based on the use of constraints technology in the *SMIL Media Adapter* to produce an updated *MModel* representation of the service. Following the user interaction in the user interface, the service listener

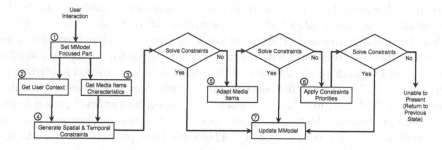

Fig. 3. Algorithm to produce an updated *MModel*

sets the focused part of the *MModel* (Step 1). Then, using the Parlay/OSA APIs, the *SMIL Media Adapter* obtains the user context description (Step 2) and extracts from the *MModel* focused part the characteristics of the media items that belong to the user interface (Step 3). These media items characteristics are the width and height of visual items and the duration of all media items.

The information collected in steps 2 and 3 enables the *SMIL Media Adapter* to establish the qualitative spatial and temporal constraints that apply to the present user situation (Step 4), which are based on the Allen's relations [20]. Applying constraint handling rules, these qualitative constraints are translated into numeric constraints that can be solved by constraint-based systems.

When the internal ECL^{i}PS^e Constraint Solver Libraries find a solution for the constraints problem, the *SMIL Media Adapter* updates the *MModel* focused part (Step 7), namely with the spatial and temporal values found for each media item. However, mainly in mobile environments where the terminals have restricted capabilities, the constraints problem may not have a solution. To solve these cases, additional constraints strategies are used by the *SMIL Media Adapter* for obtaining better results in the adaptation process. For example, when there are too many pictures to be displayed in one screen, they are distributed among multiple screen displays, maintaining the rest of the interface the remaining information. The direct access to each of these displays is provided by hyperlinks. We integrate this approach in Step 4 of the algorithm.

Another problem happens when, for example, a particular picture has dimensions that exceed the screen. In this case, a possibility is to "augment" the screen dimensions, adding a scroll bar to the user interface. However, there are some mobile devices that do not support scroll bars. Before excluding this element from the presentation, the resize of the picture to a dimension that solves the constraint problem is a possible solution. The *SMIL Media Adapter* follows this adaptation strategy (Step 5), relying also on more complex data transformations, such as picture, audio and video format conversions, text to speech or speech to text translations, performed by specific *Media Adapters* that are available in the *Adaptation Systems*.

Concerning the temporal dimension, we defined, as an input to the constraints program, the maximum duration of a presentation. This value is defined following a heuristic that takes into account the number of media items in the presentation, the user profile and the subscription information.

Another strategy to enable the resolution of constraint problem inconsistencies is to associate each constraint with a priority weight (Step 6). We follow a similar approach as Freeman et al. [21], where the constraints are ordered according to their relevance. Thus, a distinction is made between *required constraints*, which must be satisfied and, for that, are associated with a higher-level priority, and *preferred constraints*, which usually are related to the user preferences and are associated with a lower-level priority. The lower-level priority constraints are deleted from the constraint problem when they create inconsistencies.

5   Experimentation and Evaluation

As a proof of the concept we have implemented a *Customer Care* service made of an *interactive tutorials* feature, where the user interactively accesses information about products and services of a given supplier, organized as tutorials. The *SMIL Media Adapter* is evaluated with respect to some broad qualitative parameters, which include the consistency of the presented information, the adaptation capability and the presentation quality. In addition, the impact of the amount of media items integrating the user interface in the *SMIL Media Adapter* behavior is analyzed. The tested parameters were the adapter average error rate and the initial presentation delay time.

The left side of Figure 4 illustrates the structure of an *interactive tutorial* presentation as it would be displayed in ideal terminal and network conditions. The presentation still parts are, at the top, the service name and the tutorial name, and, at the bottom, the navigational buttons. The presentation moving part, at the middle, is a slide show, which corresponds to the sequential display of slides. Each slide is composed by the title, a text, pictures and video clips, possibly accompanied with audio clips for the picture slides. According to different user context conditions, the above structure can change either in form, content or both.

5.1   Qualitative Evaluation

In SMIL, spatial media are placed within boxes with predefined sizes. The position of a media item is fixed in relation to the other media. Therefore, in our approach, determining the position of a media item is equivalent to finding its coordinates using a constraint solver. The data types that have the most important impact on the presentation spatial layout are pictures and video clips. The *SMIL Media Adapter* uses a Java utility that extracts the image dimensions of each image and video clip that should be part of a presentation and integrates them in the constraint program.

The text data type is also modeled as a box, whose dimensions are calculated using an algorithm that considers the number of characters, the font size and the width of the box. Thus, the width of the text box is fixed by the presentation

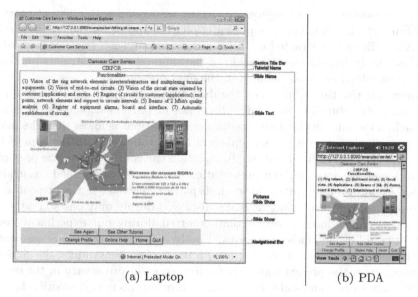

(a) Laptop | (b) PDA

Fig. 4. SMIL presentation structure of the *Interactive tutorial* feature displayed on different terminal types

structure, by other media types and by the screen dimensions, varying only the text box height.

We tested the *SMIL Media Adapter* in two terminals, a laptop and a PDA, both with the Internet Explorer browser as the SMIL player. In a PDA terminal, although the overall presentation structure shown in the left side of Figure 4 is maintained, the number of adaptation operations carried out by the *SMIL Media Adapter* is significantly higher that in large terminals. A primary adaptation operation is usually performed over text objects, adjusting the font size to an appropriate value, taking into account the small screen dimensions. Then, because the width of some pictures and video clips of the tutorial is larger than the screen width, this fact forces the *SMIL Media Adapter* to adapt them, requesting the use of an image/video resize *Media Adapter* (during Step 5 of the algorithm presented in Section 4). The resize is performed if the percentage of the reduction does not go beyond a specified threshold, usually 50% of the original size, which is defined as a limit to preserve information consistency. If a specific picture or video clip needs a resize greater than the mentioned threshold to fit in the PDA screen, the *SMIL Media Adapter* takes the decision to drop it from the tutorial presentation.

We assigned a start time and a duration to each slide displayable element. The start time of each element inside a slide is the same, except for the images, video and audio clips that can be seen as a slide show inside the main slide show. We defined, as an input to the constraint program, the maximum duration of a presentation. This value is defined following an heuristic that takes into account the user profile and the subscription information. Different profile types, such as *user@home* or *user@office*, are usually associated with different network conditions, which can impact the presentation time and consequently the service cost. Also different contracted qualities of service provision (e.g., *gold* or *silver* qualities) surely impact the definition of the time-related constraints and even the content selection to be included in the final presentation.

Concerning the time dimension, the data that have more impact in the presentation global duration are audio clips and, in particular, speech. In a first approach, if the sum of the audio media items durations is lower than the maximum presentation duration, each slide will last a time equal to the sum of the correspondent audio clips duration. To maintain the consistency of the presentation and guarantee a minimum presentation quality, we established constraints that force a minimum duration for each displayable element in the presentation, to avoid situations such as a text being displayed during only one second.

The constraints inconsistencies that occurred during our experiments were due to two different kinds of reasons. Firstly, when the durations associated with the audio clips are not consistent with the present constraints, the clips are eliminated from the presentation. This causes some inconsistency in the overall presentation, since some slides have audio and others do not. A solution for this problem is not easy to find and will always be service dependent.

The other source of constraints inconsistencies had to do with the media minimum duration constraints. These inconsistencies occurred in slides that had associated a large number of pictures, which led to a very short picture duration. In this case, the *SMIL Media Adapter* decides to drop a number of pictures from a slide, eliminating the constraints inconsistency.

5.2   Performance Evaluation

In order to evaluate the efficiency of the *SMIL Media Adapter*, a series of tests have been carried out. The objective was to analyze the behavior of the *SMIL Media Adapter* when faced with an increase of media items in a presentation. We generated a tutorial that always contained the same contents except for the number of pictures, audio and video clips of the slide show. Firstly, in the following subsection, we analyze the performance of the *SMIL Media Adapter* when the number of one media type varies. Then, the performance is analyzed in scenarios where we varied the number of two media types in a presentation.

The results shown in both subsections compare the behavior of the *SMIL Media Adapter* for four distinct levels of adaptation, closely related to the constraints algorithm presented in Section 4:

- **no adaptation**, i.e., the *SMIL Media Adapter* only converts the tutorial *MModel* to SMIL code, without applying the constraints theory, and consequently without taking into account the user context. In this case, the *SMIL Media Adapter* only applies Step 1 of the constraints algorithm;
- **with adaptation**, i.e., the *SMIL Media Adapter* generates the basic temporal and spatial constraints (Step 4), taking into account the user context and the tutorial contents, and solves them;

- **plus media adaptation**, i.e., when Step 4 of the algorithm produces a set of constraints that has no solution, the *SMIL Media Adapter* applies the media adaptation step of the algorithm (Step 5);
- **plus constraints priorities**, i.e., when the media adaptation step still produces a set of unsolved constraints, the *SMIL Media Adapter* applies constraints priorities (Step 6), enabling the deletion of low priority constraints.

Single Media Type Analysis

Figure 5 compares the impact of the slide show number of pictures on the *SMIL Media Adapter* error rate (i.e., the percentage of times the constraints problem does not have a solution) for the four levels of adaptation presented above. We considered slide shows containing a single picture up to slide shows containing 30 pictures. For each number of pictures, 50 tests were performed. The pictures were randomly picked from a database containing around 600 pictures of various dimensions. The smallest picture had 120×90 pixels, having the largest 1200×860 pixels.

The left side of Figure 5 shows the results when the tutorial is displayed in a laptop, while the right side presents the equivalent results considering a PDA as the accessing terminal. As expected, the *SMIL Media Adapter* performance decreases with the increase of the number of pictures in the slide show. This tendency is more notorious in smaller terminals, where the presentation average error rate increases in the first half of the graphic. However, in both terminals, each of the above adaptation levels contributes to a better performance of the *SMIL Media Adapter*, having a significant impact in the PDA tests, where the application of the basic constraints and the media adaptation steps are responsible for the display of the tutorial in a large number of cases. The importance of the proposed constraints algorithm in the *SMIL Media Adapter* performance is clear in the PDA graph. Here we can see that, for a large number of pictures (above 20), the basic temporal and spacial constraints step (Step 4) does not introduce

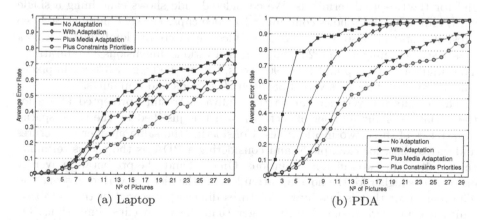

(a) Laptop   (b) PDA

Fig. 5. Presentation average error rate according to the number of pictures

Table 1. Initial presentation delay times according to the number of pictures

Number of Pictures	Laptop		PDA	
	Mean Delay	Maximum Delay	Mean Delay	Maximum Delay
4	0.2 s	1.2 s	1 s	8 s
8	0.6 s	2.2 s	5 s	12 s
12	2 s	10 s	7 s	18 s
20	6 s	18 s	10 s	25 s

any difference in the adapter performance. For these cases, the two last algorithm steps (media adaptation and constraints priorities) offer much better results.

Using the same test ambient defined above, Table 1 shows the variation of the average initial delay presentation time and the maximum initial delay presentation time of the *SMIL Media Adapter*, according to the number of pictures in the tutorial slide show. Comparing the times measured in PDAs with the laptop times, they are significantly higher. This is mainly due to the media adaptation step that starts to be used intensively for slide shows with 10 or more pictures. However, this delay is due to the media adaptation operation itself, since when the *SMIL Media Adapter* decides to perform the media adaptation over a specific tutorial element, it can, in parallel, request the constraints problem resolution, running both processes simultaneously, possibly in different machines.

The initial delay times reported in Table 1 are considerably high and unacceptable for most users. For real-world deployment, such initial delay times can be estimated, leading to the creation of new constraint priorities (e.g., do not show images larger than a predefined dimension) and their integration in the constraints problem.

Figure 6 compares the impact of the slide show number of video clips on the *SMIL Media Adapter* error rate for the four levels of adaptation presented above and for the two used terminals. We considered slide shows containing a single video clip up to slide shows containing 10 video clips. For each number of video clips, 50 tests were also performed. The video clips were randomly picked from a database containing around 200 video clips of various dimensions and durations. The smallest video clip had 120×90 pixels, having the biggest 320×240 pixels. The video clips durations varied between 20 seconds and 3 minutes.

The results are very similar to the equivalent results presented above for the variation of the number of pictures, which is justified by the same spatial proprieties of these two media types. Considering the same number of pictures and video clips, for example 5, and considering a PDA as the access terminal, the results show two different situations. On one hand, the presentation average error rate is higher for the no adaptation level in the pictures variation scenarios. The reason for that is mainly the pictures dimensions stored in the database, which are in average significantly larger than the video clips dimensions. On the other hand, the presentation average error rate is superior for the other adaptation levels in the video clips variation scenarios. This is due to the fact

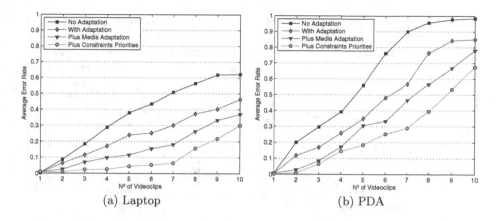

Fig. 6. Presentation average error rate according to the number of video clips

Table 2. Initial presentation delay times according to the number of video clips

Number of Video clips	Laptop		PDA	
	Mean Delay	Maximum Delay	Mean Delay	Maximum Delay
3	0.3 s	1.5 s	1.6 s	10 s
6	0.7 s	2.7 s	6 s	15 s

that the video clips have also a time dimension, which, in some tests, leads to constraints inconsistencies related to the maximum duration of a presentation.

Table 2 depicts the average initial delay presentation time and the maximum initial delay presentation time of the *SMIL Media Adapter*, according to the number of video clips in the tutorial slide show. Again, the results are similar to the pictures variation results. When the access terminal is a PDA, the initial presentation delay time is superior for video clips variation scenarios. This can be justified by the higher times required for the adaptation of video clips.

Figure 7 presents the results obtained varying the number of audio clips in the slide show. We considered slide shows containing a single audio clip up to slide shows containing 30 audio clips. 50 tests were also performed for each number of audio clips considered. The audio clips were randomly picked from a database containing around 600 clips of various durations. The smallest audio clip had 4 seconds, having the longest 4 minutes.

Audio clips have only a time dimension and the results clearly reflect this fact, in particular when the access terminal has small dimensions, as is the case of a PDA. Comparing the audio clips variation results with the results obtained with the variation of the number of pictures in a slide show presented in a PDA, for an average error rate of 50%, the number of pictures included in the presentation for the four adaptation levels are, respectively, 3, 7, 11 and 13, while the number of audio clips are 13, 15, 20 and 23. This comparison clearly shows the impact of the media spatial characteristics in the adaptation process.

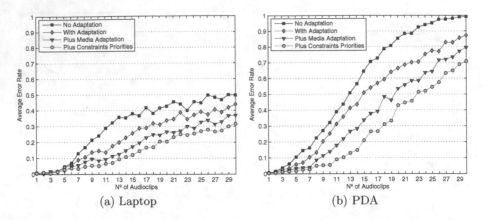

(a) Laptop   (b) PDA

Fig. 7. Presentation average error rate according to the number of audio clips

Table 3. Initial presentation delay times according to the number of audio clips

Number of Audio clips	Laptop		PDA	
	Mean Delay	Maximum Delay	Mean Delay	Maximum Delay
4	0.03 s	0.4 s	0.3 s	1.5 s
8	0.2 s	0.6 s	0.9 s	2.3 s
12	0.7 s	2 s	1.2 s	5 s
20	1 s	4 s	1.9 s	9 s

Table 3 presents the variation of the average initial delay presentation time and the maximum initial delay presentation time of the *SMIL Media Adapter*, according to the number of audio clips in the tutorial slide show. The times here are significantly lower than the times obtained with the variation of pictures and video clips, since the terminal dimensions have no direct influence in the adaptation process. The audio clips time dimension conflicts only with the maximum duration of a presentation, defined in the user profile or in the subscription contract, which is usually lower in mobile contexts than in static contexts. This explains the higher values for the initial delay presentation time obtained with PDAs.

Multiple Media Type Analysis

This section completes the analysis of the *SMIL Media Adapter* performance in respect to the space and time dimensions started in the previous section. Here, the analysis is carried out varying at the same time the number of media types that have a major impact on these dimensions. For the space dimension, we varied the number of pictures and video clips in a slide show. We used the combination of the single media type scenarios for pictures and video clips presented in the previous subsection.

Figure 8 shows the results for the four levels of adaptation, obtained when the tutorial is displayed in a laptop, while Figure 9 shows the same results when the tutorial is accessed from a PDA.

Analyzing the results, we can see that each level of adaptation contributes to a decrease of the presentation average error rate, except in PDA scenarios for the first level of adaptation with a high volume of pictures and video clips (above 15 and 4, respectively). However, the efficiency of the *SMIL Media Adapter* is more notorious when the access terminal has smaller dimensions, such as PDAs. Taking as an example a slide show of 6 pictures and 4 video clips, in PDA scenarios the presentation average error rate with no adaptation is around 90%, decreasing to 17% when all the steps of the adaptation algoritm are used, while, in laptop scenarios, the *SMIL Media Adapter* gain corresponds to a decrease of the average error rate from 33% to 5%.

Concerning the initial delay presentation time parameter, Table 4 confirms the tendency showed by Tables 1 and 2, presenting values significantly higher when the terminal is a PDA than when the terminal is a laptop. This is due to

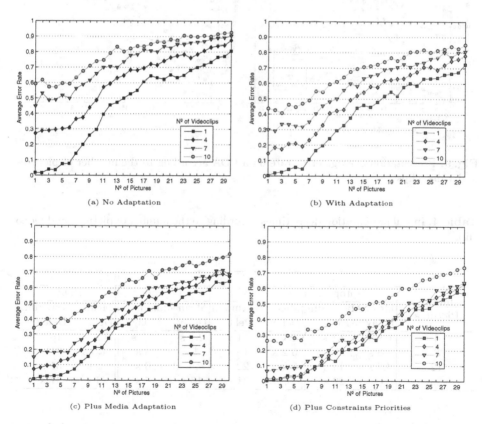

Fig. 8. Presentation average error rate on a laptop according to the number of pictures and video clips

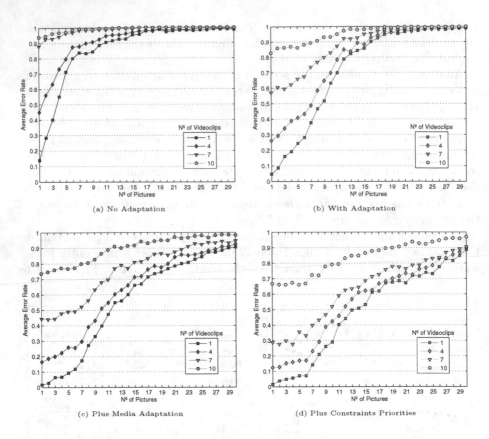

(a) No Adaptation

(b) With Adaptation

(c) Plus Media Adaptation

(d) Plus Constraints Priorities

Fig. 9. Presentation average error rate on a PDA according to the number of pictures and video clips

Table 4. Initial presentation delay times according to the number of pictures and video clips

Number of Pictures	Number of Video Clips							
	3				6			
	Laptop		PDA		Laptop		PDA	
	Mean Delay	Maximum Delay	Mean Delay	Maximum Delay	Mean Delay	Maximum Delay	Mean Delay	Maximum Delay
4	0.5 s	2.1 s	4.7 s	11 s	2.2 s	10.5 s	9.2 s	16 s
8	1.8 s	9 s	6.2 s	16 s	4 s	13 s	11.5 s	21 s
12	4.2 s	14 s	8.2 s	22 s	6.5 s	20 s	15 s	31 s
20	8 s	22 s	14 s	31 s	16 s	27 s	23 s	47 s

the fact that both media types, pictures and video clips, have the same behavior concerning the display in a terminal, i.e., both occupy space in the terminal display.

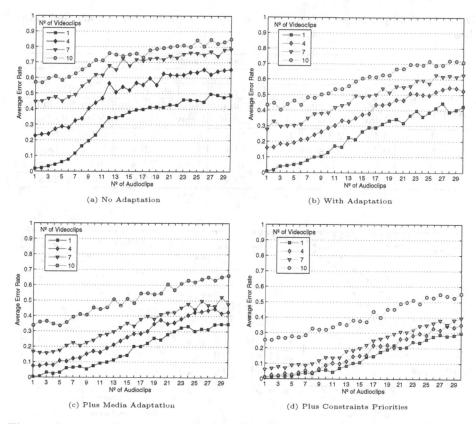

(a) No Adaptation

(b) With Adaptation

(c) Plus Media Adaptation

(d) Plus Constraints Priorities

Fig. 10. Presentation average error rate on a laptop according to the number of audio and video clips

For the time dimension analysis, we varied the number of audio and video clips in a slide show. Again, the combination of the single media type scenarios for audio and video clips, presented in the previous subsection, was used to analyze the performance of the *SMIL Media Adapter*.

Figure 10 shows the results for the four levels of adaptation, obtained when the tutorial is displayed in a laptop, while Figure 11 shows the same results when the tutorial is accessed from a PDA.

The results clearly show that the constraints algorithm decreases the presentation average error rate, in this case in all tested scenarios. Comparing these results with the spatial dimension analysis results, where the number of pictures and video clips vary, the presentation average error rate values are significantly lower. This fact is due to the reduced number of media items with spatial dimensions (10 video clips, at the maximum) for the presentations generated in these scenarios. Consequently, the gain of the constraints algorithm is not so significative as in the previous analysis. Considering an equivalent example of a slide show of 6 audio clips and 4 video clips, in PDA scenarios the presentation average error rate with no adaptation is around 47%, decreasing to 13% when

Table 5. Initial presentation delay times according to the number of audio and video clips

Number of Audio Clips	Number of Video Clips							
	3				6			
	Laptop		PDA		Laptop		PDA	
	Mean Delay	Maximum Delay	Mean Delay	Maximum Delay	Mean Delay	Maximum Delay	Mean Delay	Maximum Delay
4	0.4 s	2.2 s	1.9 s	13 s	0.8 s	3.3 s	6.8 s	16 s
8	0.7 s	2.5 s	3.1 s	14 s	1.1 s	4.9 s	7.5 s	18 s
12	1.2 s	5 s	4 s	18 s	1.9 s	7.2 s	9 s	21 s
20	1.8 s	8 s	7 s	22 s	2.7 s	12 s	11 s	29 s

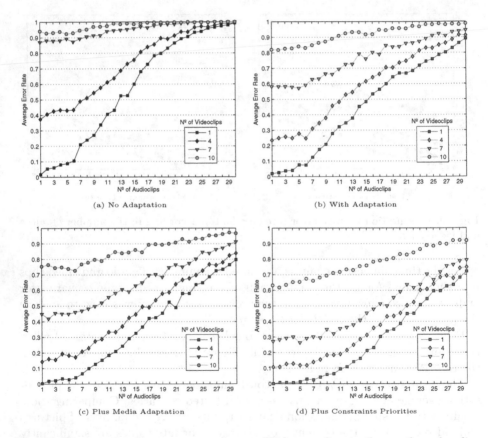

(a) No Adaptation

(b) With Adaptation

(c) Plus Media Adaptation

(d) Plus Constraints Priorities

Fig. 11. Presentation average error rate on a PDA according to the number of audio and video clips

all the steps of the constraints algorithm are used, while, in laptop scenarios, the *SMIL Media Adapter* gain corresponds to a decrease of the average error rate from 28% to 3%. These results confirm what was shown in the previous subsection, where the lower presentation average error rates were obtained in scenarios where the number of audio clips varies.

Table 5 presents the variation of the average initial delay presentation time and the maximum initial delay presentation time of the *SMIL Media Adapter*, according to the number of audio and video clips in the tutorial slide show. In general, the times here are lower than the ones presented in Table 4, showing that the audio media type has a minor impact in the adaptation process than the spatial media types (pictures and video clips).

6 Conclusions

In this article, we proposed the application of the constraints theory as the solution for defining multimedia user interface parameters dynamically adaptable to the user context. The *SMIL Media Adapter* relies on the use of the Parlay/OSA APIs, to detect, in real time, user contextual changes in mobile environments. These changes are then used to generate a set of constraints that the user interface should satisfy so that contents may be properly displayed in the current conditions. From the large number of tests carried out to evaluate the performance of the *SMIL Media Adapter*, we conclude that the proposed constraints algorithm can achieve significant improvements in the efficient adaptation of multimedia services targeted to mobile terminals.

The challenges we intend to address in future work include an user evaluation in terms of readability of adapted content, in addition to the objective assessment presented in this chapter, and also the use of real-life documents and web pages in the experiments, for comparing and validating the results obtained with a pre-established set of media items.

References

1. van Ossenbruggen, J., Geurts, J., Cornelissen, F., Hardman, L., Rutledge, L.: Towards Second and Third Generation Web-Based Multimedia. In: 10th International World Wide Web Conference, Hong Kong, pp. 479–488 (2001)
2. Scherp, A., Boll, S.: Paving the Last Mile for Multi-Channel Multimedia Presentation Generation. In: 11th International Multimedia Modelling Conference, Melbourne, Australia, pp. 190–197 (2005)
3. Bertolotti, P., Gaggi, O., Sapino, M.: Dynamic Context Adaptation in Multimedia Documents. In: ACM Symposium on Applied Computing, Dijon, France, pp. 1374–1379 (2006)
4. Ojala, T., Korhonen, J., Aittola, M., Ollila, M., Koivumäki, T., Tähtinen, J., Karjaluoto, H.: SmartRotuaari - Context-Aware Mobile Multimedia Services. In: 2nd International Conference on Mobile and Ubiquitous Multimedia, Norrköping, Sweden (2003)
5. Camarillo, G., García-Martín, M.: The 3G IP Multimedia Subsystem: Merging the Internet and the Cellular Worlds. John Wiley & Sons, Chichester (2006)
6. Phanouriou, C.: UIML: A Device-Independent User Interface Markup Language. PhD Thesis, Virginia Polytechnic Institute and State University (2000)

7. Boll, S., Klas, W.: ZYX - A Multimedia Document Model for Reuse and Adaptation of Multimedia Content. IEEE Transactions on Knowledge and Data Engineering 13(3), 361–382 (2001)
8. Müller, W., Schäfer, R., Bleul, S.: Interactive Multimodal User Interfaces for Mobile Devices. In: 37th Hawaii International Conference on System Sciences, Waikoloa, HI, USA (2004)
9. 3GPP: Multimedia Messaging Service (MMS); Media Formats and Codecs (Release 7). Technical Specification 3GPP TS 26.140 v7.1.0. Services and System Aspects Group (2007)
10. Pihkala, K., Vuorimaa, P.: Nine Methods to Extend SMIL for Multimedia Applications. Multimedia Tools and Applications 28(1), 51–67 (2006)
11. Hwang, Y., Kim, J., Seo, E.: Structure-Aware Web Transcoding for Mobile Devices. IEEE Internet Computing 7(5), 14–21 (2003)
12. Lum, W., Lau, F.: User-Centric Adaptation of Structured Web Documents for Small Devices. In: 19th International Conference on Advanced Information Networking and Applications, pp. 507–512 (2005)
13. Chen, Y., Xie, X., Ma, W., Zhang, H.: Adapting Web Pages for Small-Screen Devices. IEEE Internet Computing 9(1), 50–56 (2005)
14. Oliveira, J., Roque, R., Carrapatoso, E., Portschy, H., Hoványi, D., Berenyi, I.: Mobile Multimedia in VESPER Virtual Home Environment. In: IEEE International Conference on Multimedia and Expo., Lausanne, Switzerland (2002)
15. Parlay Group: Parlay Specifications (2003),
   http://www.parlay.org/en/specifications/
16. W3C: Synchronized Multimedia Integration Language (SMIL 2.1) Specification. Recommendation REC-SMIL2-20051213, Synchronized Multimedia Working Group (2005)
17. Hower, W., Graf, W.: A bibliographical survey of constraint-based approaches to CAD, graphics, layout, visualization, and related topics. Knowledge-Based Systems 9(7), 449–464 (1996)
18. Gajos, K., Weld, D.: SUPPLE: Automatically Generating User Interfaces. In: International Conference on Intelligent User Interfaces (IUI 2004), Funchal, Madeira, Portugal (2004)
19. Apt, K., Wallace, M.: Constraint Logic Programming using ECLiPSe. Cambridge University Press, Cambridge (2006)
20. Allen, J.: Maintaining Knowledge about Temporal Intervals. Communications of the ACM 26(11), 832–843 (1983)
21. Freeman-Brenson, B., Maloney, J., Borning, A.: An Incremental Constraint Solver. Communications of the ACM 33(1), 54–63 (1990)

Mobile Video Surveillance Systems: An Architectural Overview

Rita Cucchiara and Giovanni Gualdi

Dipartimento di Ingengeria dell'Informazione
University of Modena and Reggio Emilia
Modena, Italy
{rita.cucchiara, giovanni.gualdi}@unimore.it

Abstract. The term *mobile* is now added to most of computer based systems as synonymous of several different concepts, ranging on ubiquitousness, wireless connection, portability, and so on. In a similar manner, also the name *mobile video surveillance* is spreading, even though it is often misinterpreted with just limited views of it, such as front-end mobile monitoring, wireless video streaming, moving cameras, distributed systems. This chapter presents an overview of mobile video surveillance systems, focusing in particular on architectural aspects (sensors, functional units and sink modules). A short survey of the state of the art is presented. The chapter will also tackle some problems of video streaming and video tracking specifically designed and optimized for mobile video surveillance systems, giving an idea of the best results that can be achieved in these two foundation layers.

Keywords: mobile surveillance, ubiquitous computer vision, video safety and surveillance, video streaming for surveillance, remote computer vision.

1 Introduction

Automatic or semi-automatic surveillance systems are typically made of several sensors, of analysis units (to obtain useful and usable outputs from the sensors), and eventually of actuators (to control or to act over the environment). This last part is often enriched with human-machine interfaces to control sensors and actuators. The sensors and the analysis units can be deployed with several degrees of mobility.

The present work is focused on *Mobile Video Surveillance* systems from an architectural point of view: as a matter of fact, the term 'mobile' introduces some fuzziness to the idea of video surveillance and it is often used in misleading ways (e.g. confusing *mobile* with *remote*). Therefore a thorough overview of the main architectural components (or modules) and features of surveillance

X. Jiang, M.Y. Ma, and C.W. Chen (Eds.): WMMP 2008, LNCS 5960, pp. 89–109, 2010.

systems will be presented, underlining the mobility-related aspects of new generations of surveillance systems. In fact, this chapter does not aim to analyze the many processing functionalities of the video surveillance systems in moving contexts: there is already a wide literature on the potential capabilities in terms of knowledge extraction and scene understanding (a comprehensive book is [1], and remarkable surveys are [2,3]).

A remarkable advantage of mobile surveillance systems is to provide surveillance and remote monitoring where and when is required, without the need to install fixed systems in specific locations. This requirement is particularly important for police and public security officers which exploit surveillance systems as a dynamic and real-time support to the investigations. It is well known, as remarked in [3], that fixed installations of surveillance systems can be a deterrent element for potential crime but often they only generate a drift of criminal behaviors to neighboring areas which might not be under surveillance. For this reason, for instance, Modena municipality will soon deploy a new mobile platform for mobile surveillance tasks with Emilia Romagna region fundings.

In the first part of the chapter a short state of the art of mobility related solutions is provided. Then the second part will be focused on two specific features of mobile surveillance: optimized video streaming for mobile surveillance and video object tracking in mobile and moving scenarios. Some experiments with a prototype called MOSES (MObile Streaming for vidEo Surveillance, [4]) will be described.

2   Mobile Surveillance Systems: The Modules and Their Features

Surveillance systems are made of several modules, and each is characterized by a set of features.

2.1   The Modules of a Surveillance System

The modules of surveillance systems can be grouped in three main branches: source, functional and sink modules (Fig. 1):

- **Source Modules:** due to the multi-sensory nature of the events to capture, surveillance systems can be equipped with sensors belonging to different domains; video data is a typical module, but it could be enriched with audio sensors, ID sensors, (e.g. biometric sensors like fingerprint, RFIDs, bar codes, etc.), or context sensors (e.g. temperature, PIR, etc.). The source module is analog, being analog the nature of the sensed events, and it is followed by an analog-to-digital converter, in order to perform digital signal processing. Even if the modern sensors generally integrate the analog-to-digital unit inside the sensor device itself, any operation from the digital conversion onward is not to be accounted as part of the source module, but it is typically a part of the second module of functional processing;

- **Functional Modules:** the functional modules perform a processing task on the input, providing an output on the same or on a different domain; they can be divided in two further parts:

  - *Signal Processing Modules:* they aim to modify or transform the properties and the quality of the input, without necessarily extracting knowledge from it. In this class we include analog-to-digital converters, as well as signal encoders and decoders, streamers, but also some sensor pre-processing; in case of video, the pre-processing is typically made of low-level and context-blind operations, such as quality enhancement (SNR), pixel-wise operations (e.g. filtering, morphology, etc.) and region-of-interest operations (resizing, cropping, etc.): the pre-processing precedes analysis of higher complexity (e.g. video analytics), and it can be directly connected to the camera.

  - *Purposive Modules:* these modules are devoted to extract information and thus knowledge from video; the commercial software, often called 'video analytics', and mainly based on computer vision and pattern recognition techniques, ranges from the most basic motion detection, up to the new generation of surveillance systems that will provide, in the (coming) future, event detection, behavioral analysis, situation assessment and automatic attention focusing.

- **Sink Modules:** the sink modules are characterized by the fact that have inputs but do not have outputs to other modules (e.g. a user front end) or, in the case they do have an output, it runs into a different sub-part of the system, such as an alarming module output would be connected to some alarm actuators or human-machine interface.

2.2   The Features of a Surveillance System

From the point of view of mobility, video surveillance systems can be classified according to the features of their modules, namely distribution, mobility and degree of motion of the sensors:

- **degree of distribution of the system**, i.e. what the physical distance among modules is. A typical distinction is:

  - *monolithic systems*: all the modules are physically tightly coupled, from the sources to the sinks, including also the functional modules. In this type of systems, the network connectivity does not play any functional role, being used at most to control the whole system. Also those multi-camera systems where the cameras, even if located at different places, are functionally connected to a monolithic system (e.g. wide-baseline stereo pairs) are considered part to this group;

  - *distributed systems*: there is physical distance among the modules, with variable degree of distribution: it might be that only one portion of the system is detached from the whole rest (e.g. the video source and grabbing of a distributed network camera system), or all the parts of the system are scattered far apart from all the others (e.g. a totally distributed system). The network connectivity in these systems plays a dominant role, and the three properties to be aware of or to keep monitored are bandwidth, latency and robustness on errors. Typically the distributed surveillance systems are also defined as *remote surveillance systems*. Indeed, the survey [2] points to the remote surveillance as one of the most challenging intelligent surveillance scenario, focusing on the problems of network bandwidth and security.

This distinction is not only architectural but directly affects the system functionalities: monolithic systems with respect to distributed ones are obviously less expensive, smaller, simpler and the direct connection with the source modules makes video data available at high resolutions and frame rates; on the opposite such systems have a limited capability of covering wide areas of surveillance and a poor flexibility in processing parallelism. Conversely, distributed systems are potentially more effective: sensors can be displaced over wide spaces and computing resources can be organized in a flexible manner. This added value is obtained at the cost of a higher architectural complexity.

– **degree of mobility of the system**, i.e. how does the location of the modules changes over time; the distinction is:

  • *fixed modules*: the location of all the modules of the system never changes. Typically, these systems are powered through external outlets and wired connected. From an algorithmic point of view, the video analytic modules can take advantage from this condition, since parameters

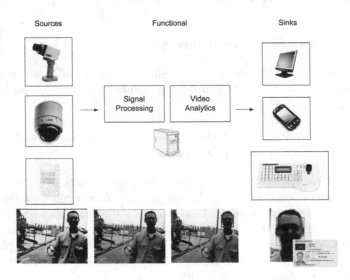

Fig. 1. Sources, functional modules and sink modules in a video surveillance system: an example for face recognition

and models can be calculated once and then repeatedly exploited thanks to the lack of ego-motion (e.g. background suppression, camera geometry and camera color calibrations);

- *mobile modules*: the modules (or a part of them) are not constrained to stay at all time in the same location. Typically, these mobile blocks are provided with wireless network connectivity, even if there are counter-examples (e.g. [5] describes a cable-based mobile sensor architecture). From an algorithmic point of view, in case the mobile part is the video source, the analytic modules will require a bigger effort to obtain useful information from the sensed environment, since parameters and models might need a re-computation after every module movement;

- *moving modules*: the physical location of the modules (or a part of them) is not only unconstrained over time, but it is also varying while the module itself is operative.

It is important to stress here the distinction between *mobile* and *moving*: the mobile module is free to move but operates while it is static, the moving module operates in motion. An example of mobile source module would be the set of cameras that observe a road construction site which moves from time to time: the cameras are moved together with the site, but they are operated being in a fixed position. An example of moving source module is camera installed over a vehicle for traffic patrolling and automatic license plate recognition.

The degree of mobility of the system has a strong impact on architectural properties of the surveillance systems such as the type of electric powering, data connection and processing architecture. The higher is the degree of mobility, the more the system will head toward battery powered solutions (rather than external outlet), small, embedded or specific purpose processor architectures (rather than general purpose ones), wireless network connectivity (rather than wired). In addition, the degree of mobility strongly affects the type of algorithms. Many video processing tasks cannot be performed with moving modules for the lack of geometric references, the reduced computational power of mobile processors and for the intrinsic different nature of the sensed data.

- **degree of movement of the (optical) sensor**, with respect to the body of the (camera) sensor (not necessarily with respect to the sensed context). There are basically two categories: *static optical sensors* (regardless of the type of optics, that can be rectilinear, omni-directional, etc.) and *movable optical sensors*, known as PTZ (pan-tilt-zoom) cameras. A new kind of sensor, that belongs to the first group even if the name could generate misconceptions, is the so called EPTZ (Electronic PTZ, [6]), that exploit high definition sensors (several mega pixels) in order to virtually reproduce the behavior of pan, tilt and zoom even if the camera and its focal length are static. There is another common misconception that confuses the mobility of the whole camera body and the mobility of its optical sensor: a clarifying counter-example is the case of a camera attached to the body of a vehicle:

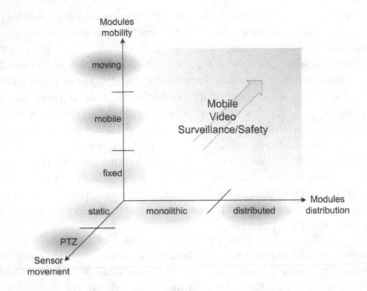

Fig. 2. The three features of a system for video surveillance or safety

because of the vehicle motion, the camera might be moving with respect to the context, but its sensor is static with respect to the camera body.

These three features are reciprocally orthogonal, as Fig. 2 represents. The systems for mobile (video) surveillance and safety are those systems that are characterized by a not-null degree of distribution, of mobility or both. There is no dependence on the third feature.

The structures of typical video surveillance systems are depicted in Fig. 3: Fig. 3(a) shows a potential monolithic system, Fig. 3(b) a distributed one, where each dashed block can be fixed, mobile or moving.

3   Towards Mobile Video Surveillance: Systems, Applications and Algorithms

This section will give an overview of the most remarkable works that directly marked the development of modern mobile video surveillance or that opened its way with preparatory findings.

Monolithic, Fixed

This is the basic architecture of the classical video surveillance systems, lacking of any degree of mobility. As depicted in Fig. 3(a), the source module sends data to a cascade of functional modules (from pre-processing to video analytics) which exploit computer vision techniques. Detected objects and events of interest are sent to the sink modules that generate alarms or annotate the videos and store

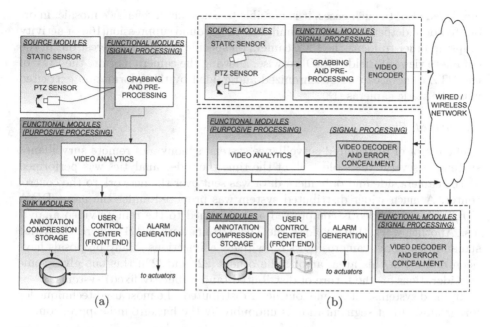

Fig. 3. Video Surveillance Systems: structure of a monolithic (a) and a distributed (b) example

them for future tasks of posterity logging. This basic architecture is well established from several years: among the first important experimentations, we find the VSAM project [7] financed by Darpa. As we will see in next sections, the VSAM project has expanded from a fixed system to a very generic mobile and distributed system. The videos, the annotations and the alarms are sent to the connected user control center front end. Being the cameras fixed, a reference model of the scene can be inferred or reconstructed to exploit background suppression for segmenting moving objects; the famous proposal of Stauffer and Grimson [8] uses Mixture of Gaussians for background modeling and suppression; a simplified method was proposed in W4 [9]. After these seminal works, tens of modified solutions have been proposed in order to account for shadows [10], multi-layer motion segmentation [11] and many others. After segmentation, discriminative appearance-based tracking is often adopted [2,12,13,14], where the appearance (color, texture, etc.) of segmented data is matched against the previously segmented and tracked objects. Still inside the monolithic systems, but closer to the border with distributed ones, there are video surveillance systems based on multiple cameras that are typically wire-connected to a central video grabber. In such systems the labels of the objects (i.e. the identity) are to be kept consistent not just along the time but also across the camera views. An example of approach for consistent labeling with overlapping views is [15], where an auto calibration is adopted and bayesian inference allows the discrimination over uncertainty due to overlapped views of groups of people. In traditional fixed systems, on top of static cameras, PTZ cameras are employed also. In such cases, it is possible to construct a reference mosaic, in order to adopt detection methods similar to background suppression[16], or activity maps, in order to focus the PTZ camera on specific regions of interest out of the complete viewable field [17]. Otherwise, knowing the camera motion, the FOV of the PTZ camera can be synchronized and guided by the information extracted with the static cameras [18].

Distributed, Fixed

The concept of distributed surveillance system, synonym of remote surveillance system, is several years old [19]: at the time the video analytics was performed at the camera side and only meta-data was sent over the network to the control center. A much wider distributed system was proposed under the name DIVA [20], where the video is compressed and streamed to the remote video analytics. Another work that exploits the DIVA framework is [21] where the video streaming of some cameras flows on wireless connections. Many proposals have been defined in this area, among the others [22] and [23]: this one gives some considerations on the future of distributed (and implicitly fixed) systems.

In fixed systems, either monolithic or distributed, the most used technique for foreground object segmentation is undoubtedly the background suppression.

Monolithic, Mobile

There is not much difference between these systems and the *monolithic, fixed* ones: the only real difference is over the video analytics algorithms that need to be working in mobile contexts. Indeed the structure of Fig. 3(a) is correct also in this scenario, with the addition that the whole block is installed on a mobile platform. The computer vision techniques that exploit a priori knowledge of the context and the scene (e.g. techniques based on the scene geometry or on the background image), can be used in this mobile contexts only if they are given the ability to be properly re-initialized or updated. A nice example of geometry recovery used for people detection and counting, that perfectly fits mobile scenarios, is the one depicted in [24], where it is possible to re-compute the geometry calibration parameters of the whole systems using a semi-supervised (and very quick) approach.

Distributed, Mobile

A distributed and mobile system often requires the introduction of wireless network communications, introducing limitations in the bandwidths: this can pose serious problems to video analytics, in case they are performed *after* the compression; [25] addresses tracking on low frame rate videos, well suited for compression that reduces temporal resolution. On the side of compression that degrades video quality, the MOSES project [4] discusses performance evaluation of mobile video surveillance applications in public parks (a brief overview of the results will be given in paragraph 5.1).

Monolithic, Moving

Those stand-alone systems that offer an unconstrained mobility, like portable and hand-held systems (e.g. smart-phones, laptops, etc.) usually belong to this category; computer vision applications limited to be fully functional within the computational borders of this kind of platforms were considered unfeasible up to few years ago, but are now spreading and there is actually a strong research activity in this area. [26] tackles face detection over PDAs, dealing with the problems of limited computational power and power consumption; [27] uses computer vision techniques over a smart phone for the visually impaired, but it does not hinders the applicability to visual surveillance. Regarding the video analytics over the monolithic moving systems, the segmentation of moving objects is completely different from what described up to this point. Frame differencing becomes useless in most of the cases (background-foreground separation is still possible though, but needs further processing: [28] is an example of foreground segmentation using KDE techniques), and detection-by-tracking instead of tracking-by-detection is often adopted: after a selection of the region of interest, several probabilistic tracking methods can be used, such as Kalman filter, mean-shift, particle filter etc [12]. An approach based on contours tracking that perfectly fits monolithic mobile systems is described in [29]. [30] proposes a unified framework for handling tracking with occlusions, non-overlapping sensor gaps and under a moving sensor that switches fields of regard. All the augmented reality works based on the SLAM [31] are an excellent example on the state of the art of what can be done on geometry recovery in monolithic moving systems.

Distributed, Moving

Given the wide complexity of this last sub-portion of the mobile video surveillance systems, we divide it in several branches:

- *moving sources, fixed sinks*: robot-surveyor applications for assessment in disaster or extraordinary scenarios were studied even in the 1980s: for example [32] deals with remote recognition of nuclear power-plants with human-based remote optical analysis; these applications did not employ any analytic module because of the limited computational capabilities of the processors at that time. In modern applications, depending on the position of the video analytics, it is possible to depict three very different cases:

  - *the video analytics is on the source side*, that is moving, and typically it is constrained in the physical dimensions, power consumption and computational power; on the opposite, the advantages are: process uncompressed video data, (possibly) stream over the network only meta-data, offer a scalable system under the computer vision point of view (the system can easily handle the increase of video sources, since each one comes with its processing unit). This is the approach based on *smart cameras*:

an interesting platform of smart cameras for video surveillance, equipped with grabbing, pre-processing, encoding and wireless radio mobile network card, is described in [33]; also [34] shows a similar example of smart cameras, mounted on autonomous robots. As we mentioned at the beginning of the chapter, VSAM [7] belongs to this category also: it streams over wireless network only 'symbolic data extracted from video signals', that is, meta-data. Actually, VSAM is a very wide project and deploys source modules of the three kinds: fixed, mobile and moving (airborne).

- *the video analytics is on the sink side*, that is fixed: typically the computation can rely on fixed and unconstrained processing platforms, and systems with mobile sources that are based on very demanding computer vision techniques, need to rely on this kind of set up. A few examples are the tracking of [35,36]. The approach of [36] will be briefly described in paragraph 5.2. The critical part is on the network links, used to transmit the whole video from the sources to functional modules: the (often limited) bandwidth must support video compression with a quality that should be sufficient for accomplishing video processing. In some cases it is possible to adopt compression schemes to improve compression efficacy (as in the case of airborne videos [37]); the recent increase in bandwidth of wireless networks (WiMAX is just an example) opens the way to new scenarios. Transmitting the whole video, and not just meta data, has the advantage that video itself can be used also for other purposes; for instance, storage or human based / human assisted surveillance cannot be performed when just meta data is streamed over the network. In case video applications do not need the streaming of the whole video (e.g. remote face recognition, [38]), even limited bandwidths can suffice. In any case, the set-up with processing positioned at the sink side looses scalability, since there is a central processing unit.
- *the video analytic is split*, partially over the source and the rest over the sinks. This can be a solution for some specific cases where it is possible to operate a clear division of functionalities inside the video analytics (e.g. low resolution processing at source side and spot-processing on high resolution images at sink side). [39] describes and analyzes the best techniques for optimized processing distribution and partition.

- *fixed sources, moving sinks*: the video analytics for this category is typically on the side of the source, that can benefit of fixed site (unconstrained physical dimensions and power supply) and uncompressed video to process. Moreover the whole system is scalable from the processing point of view. The sinks are usually reduced to front-end user monitoring and often use web-services techniques, as [40,41] depict. [42,43] study systems for video processing on fixed cameras to intelligently optimize and reduce as much as possible the transmission of video data to mobile sinks and increase perceived quality by human operators.
- *sources and sinks are independently moving*: the wireless data communication will cross two radio bridges, suffering more latency and being more prone

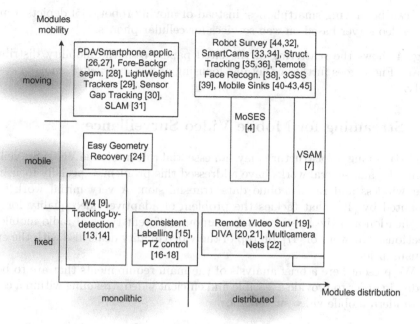

Fig. 4. Cited papers over the mobility/distribution chart

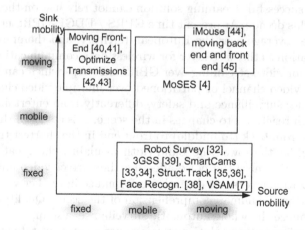

Fig. 5. Cited papers over the source/sink mobility chart

to network errors. [4] analyzes this scenario from the video streaming point of view. This is the only scenario where it is reasonable to keep the functional module fixed (and therefore detached from both sources and sinks), in order to exploit all the architectural advantages of fixed modules. iMouse [44] depicts a generic architecture based on multi-sensory surveillance where the camera sources are mounted on a moving robot and the front-end sinks can be moving smartphones; instead of moving robot, [45] depicts a mobile video server based on web-services for cellular phones.

Fig. 4 shows the scattering of the cited papers over the mobility/distribution chart. Fig. 5 goes further in details, focusing on the *distributed, moving* systems only.

4   Streaming for Mobile Video Surveillance

The streaming infrastructure plays an essential role in Mobile Video Surveillance contexts, and several works have addressed this problem, especially focusing on the wireless and radio mobile data transmission. A very initial work is represented by [46], that tackles the problem of adaptive video quality for video application and the hybrid roaming among different types of radio mobile connections. The work of [47], on top of the video quality deals also with the energy consumption.

We present here a brief analysis of the main requirements that are to be met today in order to provide a feasible and efficient video streaming within a context of modern mobile video surveillance.

- *Ubiquitousness of Connectivity*: the video coding should be sufficiently a-daptable to different wireless networks; high bandwidth radio connections such as WiFi or UMTS / HSDPA do not still offer ubiquitous coverage, therefore a successful streaming solution cannot rely just on these networks, at least in this decade. At present time GPRS / EDGE-GPRS network offers a very wide coverage over the European territory, and therefore it is taken in consideration as the base layer for wireless data communication. A typical bandwidth for video streaming over GPRS is 20kbps which can be allocated for a single video channel or partitioned over multiple video channels;
- *Latency*: since surveillance and safety, differently from entertainment, often requires high reactivity to changes in the scene, it is essential that the video is streamed from back (to middle) to front end in the shortest time possible;
- *Image Quality*: the lower the bandwidth available, the greater the video compression, the more the noise affecting the correct video analysis. Noise and quantization errors can degrade performance in further video analytics, and not just in the human comprehension of the scene. Quality degradation can compromise the whole automatic surveillance system;
- *Frame Skipping*: video analytics often require a constant rate over the video and video tracking step is a typical case where constant frame rate is usually desirable: tracking is typically based on object-to-track association on a frame basis and fixed frame rate will help in effective status predictions;
- *Frame Rate*: many algorithms of video analytics perform predictions and high frame rate can reduce the uncertainty range. In case of video tracking, since the search area for a tracked object in a new frame is generally proportional to the displacement that the object can have between two consecutive frames, a high frame rate prevents an excessively-enlarged search area.

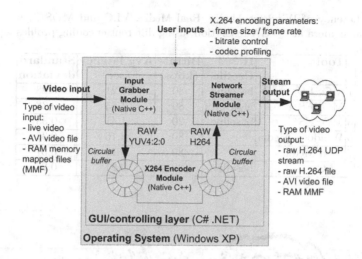

Fig. 6. Encoding layer

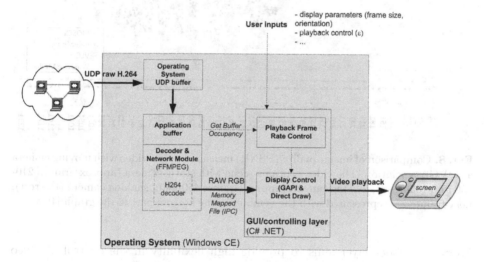

Fig. 7. Decoding layer on xScale

4.1   MOSES, a Possible Solution

MOSES (MObile Streaming for vidEo Surveillance) is an ad-hoc software implementation of a streaming system specifically designed to fulfill the afore requirements (details in [4]). It is based on H.264/AVC (MPEG-4 part 10), suitably devoted to work on low-capacity networks by means of adaptive streaming. H.264/AVC guarantees a better trade-off between image quality and bandwidth occupation with respect to MPEG-2 and MPEG-4 part 2.

The typical encoder layers of streaming systems are made of three basic blocks (video grabbing, encoding and network streaming) plus further controlling steps.

Table 1. Latency of Windows Media, Real Media, VLC and MOSES over a video with moving camera at QVGA resolution, using different encoding profiles [4]

	Tool	H.264 Profile	Bitrate (kbps)	Avg latency (sec)	Standard deviation
1	MOSES	baseline	80	1.21	0.28
4	MOSES	high prof	80	1.65	0.38
5	MOSES	baseline	20	1.41	0.33
6	Windows Media		80	4.76	0.36
7	Real Media		80	3.15	0.07
8	VLC	baseline	80	2.21	0.27

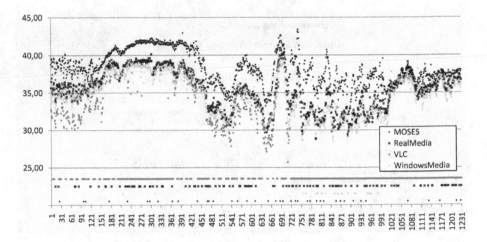

Fig. 8. Comparison of image quality (PSNR) measured over a video with moving camera at QVGA resolution. The video contains scene with static camera (approx. frames 210-450 and 1030-end), moving camera (approx. frames 450-640), shaking camera (the rest). Lost frames are represented with the symbols in the bottom part of the graph [4].

MOSES encoder layer aims to provide high flexibility in the control of video source and compression and to keep the latency and the frame-loss rate at lowest levels (Fig. 6). The following peculiar aspects of the software architecture were specifically designed in MOSES to attain such objectives: multi-threaded processing and pipeline, low buffer occupancy and UDP streaming for prioritized dispatching.

On the decoder side, MOSES deploys one module working on standard x86 platforms and another working on embedded processors (e.g. xScale): the successful implementation of mobile-architecture decoders requires peculiar optimizations, given the limited computational power of these devices. Specifically, the most critical issues are, together with video decoding and network buffering, video data exchange between processes and video display. A specific xScale based implementation is depicted in Fig. 7.

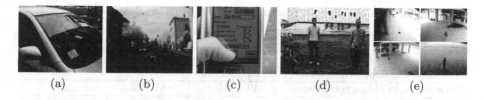

(a)   (b)   (c)   (d)   (e)

Fig. 9. Three test beds: camera-car (a,b), pda (c,d), fixed redundant cameras (e)

We tested the whole system against commercial and open streaming systems, namely Windows Media, Real Media and VideoLan, with respect to many parameters and in particular to latency, frame skipping and PSNR for image quality evaluation. In all these tests MOSES outperforms the compared systems: the better video quality and the lower latency can be attributed respectively to the use of the H.264/AVC codec and to the tightly optimized software architecture; Tab. 1 shows measurements on the latency and Fig. 8 shows image quality (measured with PSNR) and frame skipping.

We tested the streaming module from mobile devices such as a camera mounted on a car (Fig. 9 a,b), a mobile phone (c,d) and from fixed surveillance platform to mobile devices (e).

5   Video Analytics in Mobile and Moving Conditions

5.1   An Example of Video Tracking in a Scenario with Mobile Source

Video analytics algorithms working with fixed sources can be often borrowed by applications working with mobile source modules; however it becomes crucial to provide efficient re-initialization of parameters and models (e.g. background initialization) and robustness to video quality degradation due to video compression (typical in mobile scenarios). We introduce here an interesting test case over the decrease of performance of object tracking, where the mobile source modules are connected to the video analytic module by means of a low bandwidth GPRS network. The video analytic is based on SAKBOT (Statistical And Knowledge-Based Object Tracker) [10], a system for moving object detection and tracking from fixed camera. The accuracy of the overall system has been measured in terms of both pixel-level segmentation and object-level tracking, by comparing the results achieved by SAKBOT on the original, non-compressed video with those obtained on the compressed, streamed video. As Tab. 2 shows, the degradation of tracking quality due to strong video compression is not negligible, but even at extremely low bandwidths, video tracking can still provide useful results, since less then 11% of the tracks are lost or compromised using MOSES streaming in outdoor scenarios at 5kbps (for visual results see Fig. 10).

Since the compression and streaming generates frame losses, a way to align the two videos (original and compressed) for having a correct and fair comparison

Table 2. Accuracy of video tracking in mobile video surveillance scenario [4]

	# objs	Accuracy		
		ours	Real Networks	Window Media
Hall Scenario 5kbps compression	29	96.51%	81.32%	n/a
Park Scenario 5kbps compression	49	89.11%	67.95%	n/a
Park Scenario 20kbps compression	49	91.91%	91.83%	89.87%

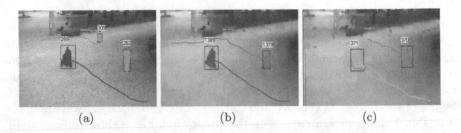

(a)   (b)   (c)

Fig. 10. Visual results of video tracking over a video scene at park: tracking on original uncompressed video (a), on video compressed at 5kbps with MOSES (b), on video compressed at 5kbps with Real Networks (c)

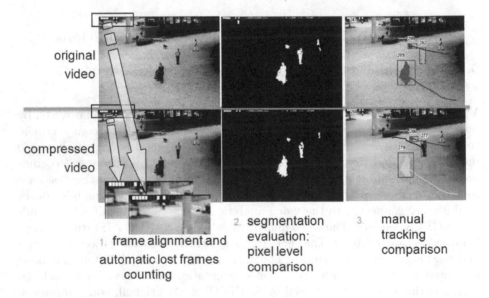

Fig. 11. Procedure used for tracking evaluation

was employed. The procedure is depicted in Fig. 11: a coded frame number on each frame of the original sequence was superimposed and such code was then used on original and on compressed videos for video alignment, latency measurement and lost frame counting.

Table 3. Summary of the performance of our graph based approach vs CamShift and Particle Filtering. Video 1 and 2 are outdoor, with freely moving camera, scene cuts, sharp zooms and camera rolls (Fig. 9(d) is taken from Video 2). Video 3 is a publicly available indoor video (AVSS 2007 dataset) with 3 people continuously occluding each other. 3-a and 3-b refer to the same video, but differ for the tracking target [36].

Obj. level	Recall			Precision			F-measure		
	CS	PF	GB	CS	PF	GB	CS	PF	GB
Video 1	96,72%	96,41%	99,24%	87,66%	74,52%	95,31%	91,97%	84,07%	97,24%
Video 2	92,83%	95,99%	100,00%	66,86%	89,00%	97,20%	77,73%	92,36%	98,58%
Video 3-a	30,25%	72,46%	97,87%	12,94%	69,89%	89,58%	18,13%	71,15%	93,54%
Video 3-b	88,59%	86,28%	98,13%	91,55%	79,39%	87,65%	90,05%	82,69%	92,59%
Pixel level	Recall			Precision			F-measure		
	CS	PF	GB	CS	PF	GB	CS	PF	GB
Video 1	84,54%	66,92%	85,71%	52,77%	49,80%	62,14%	64,09%	55,41%	71,16%
Video 2	53,20%	49,42%	84,87%	36,85%	64,64%	76,20%	43,34%	55,50%	79,78%
Video 3-a	7,93%	30,68%	65,26%	4,76%	36,93%	53,42%	5,67%	32,07%	57,67%
Video 3-b	65,06%	44,20%	71,95%	69,67%	74,30%	63,65%	64,66%	52,74%	66,90%

In conclusion, a standard surveillance module designed to work with fixed sources, might be successfully exploited also over mobile source scenarios, even when the video analytics receive highly compressed video streams.

5.2   Video Tracking in a Scenario with Moving Source

When video streams are produced from moving source modules, the intrinsic difficulty of video analytics steps up to a higher level, not only due to the low bandwidths for video compression but also to the nature of the video generated from freely moving cameras, that hinders the use of many techniques based on the prior knowledge of a background. Video taken by hand-held devices usually show a strong degree of motion, no fixed background, sharp pan, tilt and roll movements and also some frame skipping due to possible packet drops. Dealing again with a reference problem like the video tracking, kernel based tracking like meanshift [48] happens to be very appropriate for these conditions; another approach is to use particle filtering techniques such as the one proposed in [49]. Both need an initialization of the model. In case the target object is subject to occlusions/deformations or the background is strongly cluttered, the use of a traditional kernel based tracking might yield to low performances due to distractors; therefore to reinforce it we developed a pictorial structure model tracking inspired to [50]: each part of the structure is initialized on different parts of the target, and then tracked with a kernel based approach: then, the overall target tracking is maintained and guaranteed through a graph-based approach: at each new frame, in the redundant graph that contains all the potential tracked parts of the structure (one true positive and many false positives per part), the correct object tracking is extracted as the best subgraph isomorphism, with respect to the initial model graph. This search is performed using the dominant set framework [51] that is an heuristic to effectively extend the clique search over weighted graphs. The weights on the graph are measured using a energy function that maximizes the affinity of the actual observations and structure over the initial model. This graph-based approach (GB) was tested against CamShift (CS, [52]) and particle filtering (PF) on 3 videos as shown in Tab. 3, showing promising results. In-depth details can be found in [36].

6   Conclusions

The chapter has depicted what are the architectural modules and features of a generic mobile video surveillance system, showing what are the most remarkable scientific works over the years that have prepared the way or have actively contributed to its definition. The chapter then addresses the technical aspects of two foundation blocks for mobile video surveillance: video streaming and video tracking. The first one is tightly inter-connected with architectural aspects, the second is more algorithmic. The chapter shows that today both video streaming and video tracking achieve promising performances in real world scenarios, that just a few years ago would have been considered not feasible. Stable and reliable results for commercial products are probably not there yet, but this chapter shows that the research community has achieved big step forward in the last decade and that it must continue to invest in this direction.

References

1. Javed, O., Shah, M.: Automated visual surveillance: Theory and practice. The International Series in Video Computing 10 (2008)
2. Hu, W., Tan, T., Wang, L., Maybank, S.: A survey on visual surveillance of object motion and behaviors. IEEE Transaction on Systems, Man, and Cybernetics, Part C: Applications and Reviews 34 (2004)
3. Dee, H.M., Velastin, S.A.: How close are we to solving the problem of automated visual surveillance?: A review of real-world surveillance, scientific progress and evaluative mechanisms. Mach. Vision Appl. 19(5-6), 329–343 (2008)
4. Gualdi, G., Prati, A., Cucchiara, R.: Video streaming for mobile video surveillance. IEEE Transactions on Multimedia 10(6), 1142–1154 (2008)
5. Li, S., Wang, X., Li, M., Liao, X.: Using cable-based mobile sensors to assist environment surveillance. In: 14th IEEE International Conference on Parallel and Distributed Systems, pp. 623–630 (2008)
6. Bashir, F., Porikli, F.: Collaborative tracking of objects in eptz cameras. In: Visual Communications and Image Processing 2007, vol. 6508(1) (2007)
7. Collins, R., Lipton, A., Kanade, T., Fujiyoshi, H., Duggins, D., Tsin, Y., Tolliver, D., Enomoto, N., Hasegawa, O.: A system for video surveillance and monitoring. Technical Report CMU-RI-TR-00-12, Robotics Institute (2000)
8. Stauffer, C., Grimson, W.: Learning patterns of activity using real-time tracking. IEEE Trans. on PAMI 22(8), 747–757 (2000)
9. Haritaoglu, I., Harwood, D., Davis, L.S.: W4: Real-time surveillance of people and their activities. IEEE Trans. on PAMI 22, 809–830 (2000)

10. Cucchiara, R., Grana, C., Piccardi, M., Prati, A.: Detecting moving objects, ghosts and shadows in video streams. IEEE Trans. on PAMI 25(10), 1337–1342 (2003)
11. Smith, P., Drummond, T., Cipolla, R.: Layered motion segmentation and depth ordering by tracking edges. IEEE Trans. on PAMI 26(4), 479–494 (2004)
12. Yilmaz, A., Javed, O., Shah, M.: Object tracking: A survey. ACM Comput. Surv. 38(4) (2006)
13. Vezzani, R., Cucchiara, R.: Ad-hoc: Appearance driven human tracking with occlusion handling. In: Tracking Humans for the Evaluation of their Motion in Image Sequences. First Int'l Workshop (2008)
14. Senior, A., Hampapur, A., Tian, Y.L., Brown, L., Pankanti, S., Bolle, R.: Appearance models for occlusion handling. Image and Vision Computing 24(11), 1233–1243 (2006)
15. Calderara, S., Cucchiara, R., Prati, A.: Bayesian-competitive consistent labeling for people surveillance. IEEE Trans. on PAMI 30(2), 354–360 (2008)
16. Cucchiara, R., Prati, A., Vezzani, R.: Advanced video surveillance with pan tilt zoom cameras. In: Workshop on Visual Surveillance (2006)
17. Davis, J.W., Morison, A.M., Woods, D.D.: An adaptive focus-of-attention model for video surveillance and monitoring. Mach. Vision Appl. 18(1), 41–64 (2007)
18. Del Bimbo, A., Pernici, F.: Uncalibrated 3d human tracking with a ptz-camera viewing a plane. In: Proc. 3DTV International Conference: Capture, Transmission and Display of 3D Video (2008)
19. Foresti, G.: Object recognition and tracking for remote video surveillance. IEEE Transactions on Circuits and Systems for Video Technology 9(7), 1045–1062 (1999)
20. Trivedi, M., Gandhi, T., Huang, K.: Distributed interactive video arrays for event capture and enhanced situational awareness. IEEE Intelligent Systems 20(5), 58–66 (2005)
21. Kogut, G.T., Trivedi, M.M.: Maintaining the identity of multiple vehicles as they travel through a video network. In: Proceedings of IEEE Workshop on Multi-Object Tracking, p. 29 (2001)
22. Stancil, B., Zhang, C., Chen, T.: Active multicamera networks: From rendering to surveillance. IEEE Journal of Selected Topics in Signal Processing 2(4), 597–605 (2008)
23. Konrad, J.: Videopsy: dissecting visual data in space-time. IEEE Communications Magazine 45(1), 34–42 (2007)
24. Zhong, Z., Peter, V., Alan, L.: A robust human detection and tracking system using a human-model-based camera calibration. In: Proceedings of 8th Int'l Workshop on Visual Surveillance (2008)
25. Porikli, F., Tuzel, O.: Object tracking in low-frame-rate video. In: SPIE Image and Video Comm. and Processing, vol. 5685, pp. 72–79 (2005)
26. Tsagkatakis, G., Savakis, A.: Face detection in power constrained distributed environments. In: Proceedings of 1st Int'l Workshop on Mobile Multimedia Processing (2008)
27. Liu, X., Li, D.H.: Camera phone based tools for the visually impaired. In: Proceedings of 1st Int'l Workshop on Mobile Multimedia Processing (2008)
28. Leoputra, W.S., Venkatesh, S., Tan, T.: Pedestrian detection for mobile bus surveillance. In: 10th International Conference on Control, Automation, Robotics and Vision, pp. 726–732 (2008)
29. Bibby, C., Reid, I.: Robust real-time visual tracking using pixel-wise posteriors. In: Forsyth, D., Torr, P., Zisserman, A. (eds.) ECCV 2008, Part II. LNCS, vol. 5303, pp. 831–844. Springer, Heidelberg (2008)

30. Kaucic, R., Amitha Perera, A., Brooksby, G., Kaufhold, J., Hoogs, A.: A unified framework for tracking through occlusions and across sensor gaps. In: Proc. of IEEE Int'l Conference on Computer Vision and Pattern Recognition, vol. 1, pp. 990–997 (2005)
31. Klein, G., Murray, D.: Improving the agility of keyframe-based SLAM. In: Forsyth, D., Torr, P., Zisserman, A. (eds.) ECCV 2008, Part II. LNCS, vol. 5303, pp. 802–815. Springer, Heidelberg (2008)
32. Silverman, E., Simmons, R., Gelhaus, F., Lewis, J.: Surveyor: A remotely operated mobile surveillance system. In: IEEE International Conference on Robotics and Automation. Proceedings, vol. 3, pp. 1936–1940 (1986)
33. Bramberger, M., Doblander, A., Maier, A., Rinner, B., Schwabach, H.: Distributed embedded smart cameras for surveillance applications. Computer 39(2), 68 (2006)
34. Silva, D., Pereira, T., Moreira, V.: Ers-210 mobile video surveillance system. In: Portuguese Conference on Artificial intelligence, pp. 262–265 (2005)
35. Ramanan, D., Forsyth, D., Zisserman, A.: Tracking people by learning their appearance. IEEE Trans. on PAMI 29(1), 65–81 (2007)
36. Gualdi, G., Albarelli, A., Prati, A., Torsello, A., Pelillo, M., Cucchiara, R.: Using dominant sets for object tracking with freely moving camera. In: Workshop on Visual Surveillance (2008)
37. Perera, A., Collins, R., Hoogs, A.: Evaluation of compression schemes for wide area video. In: IEEE Workshop on Applied Imagery Pattern Recognition, pp. 1–6 (2008)
38. Al-Baker, O., Benlamri, R., Al-Qayedi, A.: A gprs-based remote human face identification system for handheld devices. In: IFIP International Conference on Wireless and Optical Communications Networks (2005)
39. Marcenaro, L., Oberti, F., Foresti, G., Regazzoni, C.: Distributed architectures and logical-task decomposition in multimedia surveillance systems. Proceedings of the IEEE 89(10) (2001)
40. Wang, Y.K., Wang, L.Y., Hu, Y.H.: A mobile video surveillance system with intelligent object analysis. Multimedia on Mobile Devices 6821(1) (2008)
41. Imai, Y., Hori, Y., Masuda, S.: Development and a brief evaluation of a web-based surveillance system for cellular phones and other mobile computing clients. In: Conference on Human System Interactions, pp. 526–531 (2008)
42. Raty, T., Lehikoinen, L., Bremond, F.: Scalable video transmission for a surveillance system. In: Int'l Conference on Information Technology: New Generations, pp. 1011–1016 (2008)
43. Steiger, O., Ebrahimi, T., Cavallaro, A.: Surveillance video for mobile devices. In: IEEE Conference on Advanced Video and Signal Based Surveillance, pp. 620–625 (2005)
44. Tseng, Y.C., Wang, Y.C., Cheng, K.Y., Hsieh, Y.Y.: imouse: An integrated mobile surveillance and wireless sensor system. Computer 40(6), 60–66 (2007)
45. Wu, B.F., Peng, H.Y., Chen, C.J., Chan, Y.H.: An encrypted mobile embedded surveillance system. In: Intelligent Vehicles Symposium, Proceedings, pp. 502–507. IEEE, Los Alamitos (2005)
46. Davies, N., Finney, L., Friday, A., Scott, A.: Supporting adaptive video applications in mobile environments. IEEE Communications Magazine 36(6) (1998)
47. Chiasserini, C.F., Magli, E.: Energy consumption and image quality in wireless video-surveillance networks. In: IEEE International Symposium on Personal, Indoor and Mobile Radio Communications (2002)

48. Comaniciu, D., Ramesh, V., Meer, P.: Real-time tracking of non-rigid objects using mean shift. In: Proc. of IEEE Int'l Conference on Computer Vision and Pattern Recognition, vol. 2, pp. 142–149 (2000)
49. Pérez, P., Hue, C., Vermaak, J., Gangnet, M.: Color-based probabilistic tracking. In: Heyden, A., Sparr, G., Nielsen, M., Johansen, P. (eds.) ECCV 2002. LNCS, vol. 2350, pp. 661–675. Springer, Heidelberg (2002)
50. Felzenszwalb, P.F., Huttenlocher, D.P.: Pictorial structures for object recognition. Int. J. Comput. Vision 61(1), 55–79 (2005)
51. Pavan, M., Pelillo, M.: Dominant sets and pairwise clustering. IEEE Trans. on PAMI 29(1), 167–172 (2007)
52. Bradski, G.: Real time face and object tracking as a component of a perceptual user interface. In: Proc. of IEEE Workshop on Applications of Computer Vision, pp. 214–219 (1998)

Hybrid Layered Video Encoding for Mobile Internet-Based Computer Vision and Multimedia Applications

Suchendra M. Bhandarkar[1], Siddhartha Chattopadhyay[2],
and Shiva Sandeep Garlapati[1]

[1] Department of Computer Science, The University of Georgia, Athens,
GA 30602-7404, USA
[2] Google Inc., Mountain View, CA 94043, USA

Abstract. Mobile networked environments are typically resource constrained in terms of the available bandwidth and battery capacity on mobile devices. Real-time video applications entail the analysis, storage, transmission, and rendering of video data, and are hence resource-intensive. Since the available bandwidth in the mobile Internet is constantly changing, and the battery life of a mobile video application decreases with time, it is desirable to have a video representation scheme that adapts dynamically to the available resources. A *Hybrid Layered Video* (HLV) encoding scheme is proposed, which comprises of content-aware, multi-layer encoding of texture and a generative sketch-based representation of the object outlines. Different combinations of the texture- and sketch-based representations result in distinct video states, each with a characteristic bandwidth and power consumption profile. The proposed HLV encoding scheme is shown to be effective for mobile Internet-based multimedia applications such as background subtraction, face detection, face tracking and face recognition on resource-constrained mobile devices.

Keywords: video streaming, layered video, mobile multimedia.

1 Introduction

The increasing deployment of broadband networks and simultaneous proliferation of low-cost video capturing and multimedia-enabled mobile devices, such as pocket PCs, cell phones, PDA's and iPhones during the past decade have triggered a new wave of mobile Internet-based multimedia applications. Internet-scale mobile multimedia applications such as video surveillance, video conferencing, video chatting and community-based video sharing are no longer mere research ideas, but have found their way in several practical commercial products. Most multimedia applications typically entail the analysis, transmission, storage and rendering of video data and are hence resource-intensive. Mobile network environments, on the other hand, are typically resource constrained in terms of the available bandwidth and battery capacity on the mobile devices. Mobile Internet environments are also characterized by constantly fluctuating bandwidth and decreasing device battery life

X. Jiang, M.Y. Ma, and C.W. Chen (Eds.): WMMP 2008, LNCS 5960, pp. 110–136, 2010.

as a function of time. Consequently, it is desirable to have a hierarchical or layered video encoding scheme where distinct layers have different resource consumption characteristics [32].

Traditional layered video encoding, as used by the MPEG-4 Fine Grained Scalability profile (MPEG-FGS), is based on the progressive truncation of DCT or wavelet coefficients [12]. There is a tradeoff between the bandwidth and power consumption requirements of each layer and the visual quality of the resulting video, i.e., the lower the resource requirements of a video layer, the lower the visual quality of the rendered video [12]. Note that the conventional MPEG-FGS layered encoding is based on the spectral characteristics of low-level pixel data. Consequently, in the face of resource constraints, the quality of the lower layer videos may not be adequate to enable high-level computer vision or multimedia applications. For a layered video encoding technique to enable a high-level computer vision or multimedia application, it is imperative that the video streams corresponding to the lower encoding layers contain enough high-level information to enable the application at hand while simultaneously satisfying the resource constraints imposed by the mobile network environment.

In this chapter, a *Hybrid Layered Video* (HLV) encoding scheme is proposed, which comprises of content-aware, multi-layer encoding of texture and a generative sketch-based representation of the object outlines. Different combinations of the texture- and sketch-based representations result in distinct video states, each with a characteristic bandwidth and power consumption profile. The high-level content awareness embedded within the proposed HLV encoding scheme is shown to enable high-level vision applications more naturally than the traditional layered video encoding schemes based on low-level pixel data. The proposed HLV encoding scheme is shown to be effective for mobile Internet-based computer vision and multimedia applications such as background subtraction, face detection, face tracking and face recognition on resource-constrained mobile devices.

2 Overview of Proposed Scheme

Video reception and playback on a mobile device such as a PDA, pocket-PC, multimedia-enabled mobile phone (for example, the iPhone), or a laptop PC operating in battery mode, is a resource-intensive task in terms of CPU cycles, network bandwidth and battery power [32]. Video reception and playback typically results in rapid depletion of battery power in the mobile device, regardless of whether the video is streamed from a hard drive on the device, or from a remote server. Several techniques have been proposed to reduce resource consumption during video playback on the mobile device [10], [11], [22], [26], [27]. These techniques use various hardware and software optimizations to reduce resource consumption during video reception and playback. Typically, resource savings are realized by compromising the quality of the rendered video. This tradeoff is not always desirable, since the user may have very well chosen to watch the video at its highest quality if sufficient battery power were available on the device. It is desirable to formulate and implement a multi-layer encoding of the video such that distinct layers of the video display different resource consumption characteristics. The lowest layer should consume the fewest resources during video reception, decoding and rendering. The resource consumption during video reception, decoding and rendering should increase as more layers are added to the video. Typically, fewer the available resources to receive, decode and render the video, the lower the quality of the rendered video on the mobile device [12]. In light of the aforementioned tradeoff between resource consumption and video quality, it is necessary to enhance quality of the lower video layers in order to ensure that the quality of the rendered video is acceptable.

Traditional layered video encoding, as used by the MPEG-4 Fine Grained Scalability profile (MPEG-FGS), is customized for varying bitrates, rather than adaptive resource usage. In this paper, we present the design and implementation of a novel *Hybrid Layered Video* (HLV) encoding scheme. The proposed representation is termed as "hybrid" on account of the fact that its constituent layers are a combination of standard MPEG-based video encoding and a generative sketch-based video representation. The input video stream is divided into two components: a *sketch* component and a *texture* component. The sketch component is a *Generative Sketch-based Video* (GSV) representation, where the outlines of the objects of the video are represented as curves [6]. The evolution of these curves, termed as *pixel-threads*, across the video frames is modeled explicitly in order to reduce temporal redundancy. A considerable body of work on object-based video representation using graphics overlay techniques has been presented in the literature [20], [21], [30]. These methods are based primarily on the segmentation of the video frames into regions and the subsequent representation of these regions by closed contours. A major drawback of the aforementioned contour-based representation is the fact that the complexity of the representation increases significantly with an increasing number of contours in the video frames. In contrast, the proposed GSV representation uses sparse parametric curves, instead of necessarily closed contours, to represent the outlines of objects in the video frames. This ensures that the number of graphical objects that one needs to overlay is small. In addition, whereas closed contours are capable of addressing local region-based consistency, global shape-based information may be seriously compromised. This is not so in the case of the proposed GSV representation, which ensures that the global shape is correctly represented. Although contour-based representations have been very successful in some specific applications involving low bitrate videos such as video phones [15], generic contour-based video representations for a wider class of power-constrained devices have, thus far, not been studied in detail.

The texture component in the proposed HLV encoding scheme is represented by three layers; a base layer video, an intermediate mid-layer video, and the original video. The base layer represents a very low bitrate video with very low visual quality whereas the highest layer in the HLV representation denotes the original video. The base layer video can be augmented by the object outlines that are emphasized with dark contours using the Generative Sketch-based Video (GSV) representation mentioned above. This ensures that the visual quality of the base layer is improved significantly. The visual quality of the mid-layer video is higher than that of the base layer video, but lower than that of the original video. The quality of the mid-layer video is further enhanced via high-level object-based re-rendering of the video at multiple scales of resolution. The result is termed as a *Features, Motion and Object-Enhanced Multi-Resolution* (FMOE-MR) video [5]. Note that although the visual quality of the mid-layer video is lower than that of the original video, the semantically relevant portions of the frames in the mid-layer video are highlighted by selectively rendering them at higher resolution, thus enhancing the overall viewing experience of the end user.

We show that the various video layers in the proposed HLV representation have different power consumption characteristics. Thus, the overall power consumption of an HLV-encoded video depends on the combination of layers used during decoding and rendering of the video on the mobile end user device. A direct benefit of the proposed HLV representation is that the video content can be received, decoded and rendered at different levels of resource consumption on the mobile device. The proposed HLV representation is thus extremely well suited for video streaming to power-constrained devices, such as multimedia-enabled mobile phones, PDAs, pocket-PCs and laptop computers operating in battery mode, where the available bandwidth and power for video reception and playback are typically observed to change over the duration of video reception and playback. A schematic depiction of the proposed Hybrid Layered Video (HLV) representation is given in Fig 1.

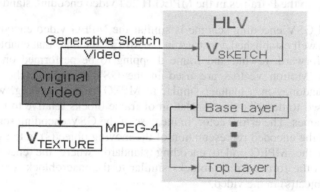

Fig. 1. The Hybrid Layered Video (HLV) Scheme

In the following sections, we elaborate upon the two components of the proposed HLV representation, i.e., the sketch component, V_{SKETCH} and the texture component, $V_{TEXTURE}$. This is followed by a description of how to combine the various video layers comprising the V_{SKETCH} and $V_{TEXTURE}$ components to arrive at a resource-scalable video representation. We discuss issues pertaining to the implementation of the proposed HLV representation followed by a presentation and analysis of the experimental results. Finally, we conclude the chapter with an outline of future research directions.

3   Creating Video Component V_{sketch}

The sketch-based video component V_{SKETCH} essentially represents the outlines of the objects in the video. We describe a technique to represent a video stream as a sequence of *sketches,* where each sketch in turn is represented by a sparse set of parametric curves. The resulting video representation is termed a *Generative Sketch-based Video* (GSV). The video is first divided into a series of *Groups of Pictures* (GOPs), in a manner similar to standard MPEG video encoding [28]. Each GOP consists of N frames (typically, $N = 15$ for standard MPEG/H.264 encoding) where each frame is encoded as follows:

1. The object outlines are extracted in each of the N frames. These outlines are represented as a sparse set of curves.
2. The curves in each of the N frames are converted to a suitable parametric representation.
3. Temporal consistency is used to remove spurious curves, which occur intermittently in consecutive frames, to remove an undesired flickering effect.
4. Finally, the parametric curves in the N frames of the GOP are encoded in a compact manner. The first frame of the GOP enumerates the curve parameters in a manner that is independent of their encoding, analogous to the I-frame in MPEG H.264 video encoding standard. The remaining N-1 frames in the GOP are encoded using motion information derived from previous frames, in a manner analogous to the P-frames in the MPEG H.264 video encoding standard.

The proposed GSV encoding scheme is similar the MPEG video encoding standard. The GOP is a well established construct in the MPEG standard that enables operations such as fast forward, rewind and frame dropping to be performed on the encoded video stream. Motion vectors are used in the GSV encoding scheme to reduce temporal redundancy in a manner similar to MPEG video encoding, where motion vectors are used to describe the translation of frame blocks relative to their positions in previous frames. The error vector, in the case of the GSV encoding scheme, has the same form as the encoded representation of the moving object(s) in the video. This is analogous to the MPEG video encoding standard, where the encoding error is represented in the form of macroblocks similar to the macroblock representation of the moving object(s) in the video.

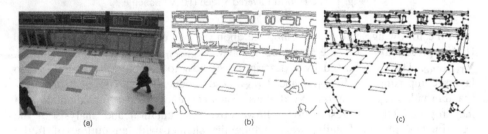

(a)   (b)   (c)

Fig. 2. The creation of *pixel-threads* for a video frame (a) The original video frame; (b) Edges detected in the video frame, and filtered to remove small, spurious edges; (c) Break-points detected in the edge contours generated in the previous step

The parametric curves used to represent the object outlines in each frame are termed as *pixel-threads*. A pixel-thread is derived from a polyline P:[0, N], which is a continuous and piecewise linear curve made of N connected segments. A polyline can be parameterized using a parameter $\alpha \in \Re$ (set of real numbers) such that P(α) refers to a specific position on the polyline, with P(0) referring to the first vertex of the polyline and P(*N*) referring to its last vertex. Note that the pixel-threads contain information only about the vertices (or break points) of the polyline. Note that these break points can be joined by straight line segments (as in the case of a polyline), or by more complex spline-based functions to create smooth curves.

The pixel-threads essentially depict the outlines of objects in the underlying video stream. Each video frame is associated with its own collection of pixel-threads termed as a *Pixel-thread-Pool*. Thus, successive video frames are associated with successive Pixel-thread-Pools. Due to the temporal nature of the video, the pixel-threads and Pixel-thread-Pools are modeled as dynamic entities that evolve over time to generate the outlines of the moving objects in the video. The dynamic nature of the pixel-threads is modeled by the processes of birth and evolution of pixel-threads over time. We provide a detailed description of these processes in the following subsections.

3.1 Birth of Pixel-Threads

For a given video frame, the *Pixel-thread-Pool* is created by first generating (or sketching) the outlines of the objects in the video frame, and then representing these outlines parametrically in the form of *pixel-threads*.

Generating a sketch from a video frame. The edge pixels in a video frame are extracted using the Canny edge detector [4]. The edge pixels are grouped to form one-pixel wide edge segments or *edgels*, many of which are intersecting. Edgels of small length are removed to avoid excessively cluttered sketches. The threshold below which an edgel is considered "small" depends on the screen size. Since GSV encoding is typically meant for mobile devices with small screens, the removal of these small edgels typically does not produce any adverse effect. It must be noted that the edge detection process is inherently sensitive to noise and several edgels may, in fact, be noisy artifacts. The edgels extracted in two successive frames may result in a *flickering* effect, wherein an edgel in a previous frame may disappear in the current frame, even in instances where the human eye can clearly discern an object boundary. A method to reduce this flickering effect is described in Section 3.4.

Creating Pixel-Threads from a Sketch. The sketch thus obtained is converted to an approximate parametric representation using curve approximation techniques proposed by Rosin and West [29]. Rosin and West [29] describe the implementation and demonstrate the performance of an algorithm for segmenting a set of connected points resulting in a combination of parametric representations such as lines, circles, ellipses, super-elliptical arcs, and higher-order polynomial curves. The algorithm is *scale invariant* (i.e., it does not depend on the size of the edgels, or the size of the frame), *nonparametric* (i.e., it does not depend on predefined parameters), *general purpose* (i.e., it works on any general distribution of pixels depicting object outlines in any given video frame), and *efficient* (i.e., has low computational time complexity). Since a detailed discussion of the algorithm is beyond the scope of the paper, it suffices to mention that we use this algorithm to determine break points on the various connected components (i.e., edge segments) that are generated after the edge pixels have been detected.

A curved edge segment is represented by a series of break points along the curve, determined using the algorithm of Rosin and West [29]. The curved edge segment is deemed to represent a portion of the outline of an object in the scene. Thus, the fitting of straight lines between the break points results in a rendering of an *approximate* version of the original curve. The break points are essentially points of significance along the curve, such as corners and high-curvature points. Altering the level of significance or threshold for break-point detection allows various levels of (break-point based) approximation of the contours in the video frame.

The break points along the curve are represented efficiently as a chain-coded vector. For each approximated curve i, one of the end points (first or last break point) is represented using absolute coordinates $\{x_0, y_0\}$ whereas the pth break point, where $p > 0$, is represented by coordinates relative to those of the previous break point; i.e. $\{\delta x_p, \delta y_p\}$ where $\delta x_p = x_p - x_{p-1}$ and $\delta y_p = y_p - y_{p-1}$. The resulting chain-coded vectors constitute the *pixel-threads* which are approximations to the original curve. Fig. 2 illustrates the process by which *pixel-threads* are generated for a given video frame. Note that the *pixel-thread* creation procedure is performed *offline* during the encoding process.

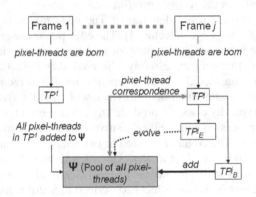

Fig. 3. Illustration of the process of establishing *pixel-thread* correspondence for a frame j and the current *All-Threads-Pool* Ψ

3.2 Evolution of a Pixel-Thread

Due to the temporal redundancy in a video sequence, a majority of the corresponding *pixel-threads* in successive video frames are often similar in shape and size. This temporal redundancy can be exploited to *evolve* some of the *pixel-threads* in the current frame to constitute the *Pixel-thread-Pool* for the successive frame. An advantage of evolving the *pixel-threads* over successive video frames is the fact that a *pixel-thread*, once *born* in a video frame, requires only motion information to characterize its behavior in successive frames. Motion modeling significantly reduces the amount of information required to render the set of *pixel-threads* belonging to the next frame. This results in a compact representation of the dynamic *pixel-threads*. The evolution parameters are determined during the encoding process, which is performed offline and typically not resource constrained. As will be shown subsequently, the encoding is done in a manner such that the decoding is simple, and can be done in real time by a resource constrained device.

The evolution of *pixel-threads* between two successive *Pixel-thread-Pools,* say *TP₁* and *TP₂,* involves two steps; (a) establishing the *pixel-thread* correspondence between the two *Pixel-thread-Pools,* and (b) estimating the motion parameters.

Establishing Pixel-Thread Correspondence. In order to model the underlying motion accurately, it is essential to establish the correspondence between *pixel-threads,* belonging to the *Pixel-thread-Pools* of two consecutive frames in the video stream. In order to determine the correspondence between *pixel-threads* in *TP₁* and *TP₂* one needs to determine for each *pixel-thread* in *TP₁* its counterpart in *TP₂.*

First, we need to predict a position to which a *pixel-thread* $T_1 \in TP_I$ is *expected* to move in the next frame. The predicted *pixel-thread,* say T', can be determined using a suitable optical flow function *OpF*( ), such that

$$T' = OpF(T_1) \tag{1}$$

The function *OpF*( ) computes the coordinates of the break points of the *pixel-thread* $T' \in TP_2$, given the coordinates of the break points of *pixel-thread* $T_1 \in TP_I$. The function *OpF*( ) implements a sparse iterative version of the Lucas-Kanade optical flow algorithm designed for pyramidal (or multiscale) computation [3]. The Lucas-Kanade algorithm is a popular version of a two-frame differential technique for motion estimation (also termed as optical flow estimation). For each break point location (x, y) of a *pixel-thread,* if the corresponding pixel location in the original image (frame) has intensity $I(x, y)$; and is assumed to have moved by δx and δy between the two frames, then the image constraint equation is given by:

$$I_{current-frame}(x, y) = I_{next-frame}(x + \delta x, y + \delta y)$$

The Lucas-Kanade algorithm essentially embodies the above image constraint equation. The pyramidal implementation of the Lucas-Kanade algorithm computes the optical flow in a coarse-to-fine iterative manner. The spatial derivatives are first computed at a coarse scale in scale space (i.e., in a pyramid), one of the images is warped by the computed deformation, and iterative updates are then computed at successively finer scales.

Once the *pixel-thread* T' is obtained from T_1 via the optical flow function, we hypothesize that if *pixel-thread* T_1 in *Pixel-thread-Pool TP₁* does indeed evolve to a corresponding *pixel-thread* T_2 in *TP₂,* then T' and T_2 should resemble each other (to a reasonable extent) in terms of shape and size. The key is to determine the *pixel-thread* T_2 in *TP₂,* which is closest in shape and size to the *pixel-thread* T'.

The correspondence between pixel threads T' and T_2 is determined using the *Hausdorff distance* [1]. The Hausdorff distance is used as a measure of (dis)similarity between the *pixel-threads,* T' and T_2. The Hausdorff distance between the two *pixel-threads* T' and T_2, denoted by $\delta_H(T', T_2)$, is defined as

$$\delta_H(T', T_2) = \max_{a \in T'} \{ \min_{b \in T2} \{ d(a, b) \} \} \tag{2}$$

where **a** and **b** are the break points on the respective curves, and d(**a**, **b**) is the Euclidean distance between them. Thus, given *pixel-thread* $T_1 \in TP_1$, T_2 is essentially the *pixel-thread* in TP_2 which is most similar to T', where T', in turn, is obtained from T_1 using the optical flow mapping function $OpF(\ )$; i.e.

$$T_2 = \operatorname{argmin}\{\ \delta_H(OpF(T_1), T): T \in TP_2\} \tag{3}$$

An important observation about the computation of the Hausdorff distance δ_H is that the two *pixel-threads* under consideration, T_1 and T_2, need not have the same number of break points.

Although the *pixel-thread* $T_2 \in TP_2$ is deemed to be the closest evolved *pixel-thread* to $T_1 \in TP_1$, it might still *not* have actually evolved from T_1. As a result, we define a threshold $\epsilon > 0$, such that if $\delta_H(OpF(T_1), T_2) < \epsilon$, then we consider T_2 to have evolved from T_1; otherwise, T_1 is deemed to have become *dormant*, and $T_2 \in TP_2$ is deemed to have been *born* in TP_2 and not evolved from TP_1. We specify the threshold ϵ as an empirically determined fraction of the video frame width.

Based on the above definitions of *birth* and *evolution*, the *pixel-threads* in TP_2 can be categorized as belonging to two mutually exclusive sets, TP^{Evolve} and TP^{born}. TP^{Evolve} is the set of all *pixel-threads* in TP_2 which are evolved from some *pixel-thread* in TP_1, and TP^{born} is the set of *pixel-threads* in TP_2 which are not evolved from TP_2; in other words, these *pixel-threads* are deemed to have been born in TP_2. Fig 3 provides a schematic description of this process.

Motion Modeling of Pixel-Threads. In this section, we discuss how to encode the motion information needed to specify the evolution of a *pixel-thread* $T_1 \in TP_1$ to its counterpart $T_2 \in TP_2$ once the correspondence between the *pixel-threads* T_1 and T_2 has been determined as described in the previous subsection. The visual quality and computational efficiency of the final encoding requires accurate estimation of the motion of *pixel-thread* T_1 as it evolves into *pixel-thread* T_2. To ensure compactness of the final representation, we assume that a linear transformation L_T, specified by the translational parameters $\{t_x, t_y\}$, can be used for the purpose of motion estimation. It must be noted that transformations based on affine motion models, that incorporate additional parameters, such as rotation and scaling, can also be used. However, simple translational motion requires the least number of bytes for representation. Moreover, even if a more accurate and comprehensive motion model were to be used, an encoding error term would still need to be computed. The encoding error generated by an accurate and comprehensive motion model, although smaller in magnitude compared to that generated by a simple motion model consisting of translational parameters only, would still require approximately the same number of bytes for representation. For example, a byte would be required to represent the encoding error in both cases, whether the error is 1 pixel or 255 pixels. Thus, using the simplest motion model (consisting only of translational parameters), and keeping track of the resulting encoding error, results in a compact and efficient motion representation. Note that a similar motion model is used for motion compensation by the well established MPEG standard. Also note that the simple translational motion model is adequate when the temporal sampling rate of the video (measured in frames per second) is high enough compared to the velocities and complexities of the motions of the various objects within the video. In such cases, even complex motions between successive frames can be reasonably approximated by a motion model comprising only of translational parameters.

Thus, the estimated *pixel-thread*, $T_2^{estimated}$ is computed from T_1 by using a mapping function L_{T1}, such that

$$T_2^{estimated} = L_{T2}(T_1)$$

The linear transformation coordinates in L_{T2} can be determined by computing the mean of the displacements of each break point, where the displacement of each break point is computed using the function $OpF(\ )$ (equation (1)). Since $T_2^{estimated}$ may not align exactly point-by-point with T_2, it is necessary to compute the error between $T_2^{estimated}$ and T_2. As discussed in the previous subsection, $T_2^{estimated}$ and T_2 may not have the same number of break points. Suppose the number of break points of T_1, and hence, $T_2^{estimated}$, is n_1 and that of T_2 is n_2. In general, $n_1 \neq n_2$. Let the displacement error between $T_2^{estimated}$ and T_2, be given by the displacement vector ΔT_2. Two cases need to be considered:

Case 1: $n_2 \leq n_1$: This means that there are fewer or equal number of break points in T_2 compared to $T_2^{estimated}$. Note that, each component of ΔT_2 is a relative displacement required to move each break point of $T_2^{estimated}$ to one of the break points in T_2. Obviously, there can be multiple break points in $T_2^{estimated}$ which map to the same break point in T_2.

Case 2: $n_2 > n_1$: In this case the encoding is slightly different. The first n_1 components of ΔT_2 denote the displacements corresponding to break points in T_2 in the same order. Each of the remaining $(n_2 - n_1)$ components of ΔT_2 are now encoded as displacements from the last break point in T_2.

From the above description, it can be seen that the displacement vector ΔT_2 has $max(n\ ,\ n_2)$ components. Thus, the motion model required to evolve *pixel-thread* T_1 into *pixel-thread* T_2, is given by

$$\Theta_{T1}(T_2) = \{\ t_x, t_y, \Delta T_2\} \tag{4}$$

The motion model $\Theta_{T1}(T_2)$ essentially contains all the parameters needed to transform *pixel-thread* $T_1 \in TP_1$ to *pixel-thread* $T_2 \in TP_2$.

Let us now consider the number of bytes required to encode the motion model $\Theta_{T1}(T_2)$. The transformation parameters $\{t_x, t_y\}$ can be designed to require a byte (character) each by restricting the displacement values to lie in the range $(-127, 128)$. If t_x or t_y exceeds these bounds, then the result of the correspondence determination procedure is declared void, and T' is deemed to be a new *pixel-thread* that is born, instead of one that is evolved from *pixel-thread* T. However, in practice, for small display screens typical of mobile devices, this case occurs very rarely. The prediction error vector ΔT_2 requires 2 bytes for each component, if the displacements δx and δy are restricted to lie within a range $\{-128, 127\}$. Thus, ΔT_2 requires $2 \cdot max(n_1\ ,\ n_2)$ bytes of storage. Hence, the total storage requirement of the motion model $\Theta_{T1}(T_2)$ (in bytes), denoted by $Bytes(\Theta_{T1})$, is given by

$$Bytes(\Theta_{T1}) = 2 \cdot max(n_1\ ,\ n_2) + 2 \tag{5}$$

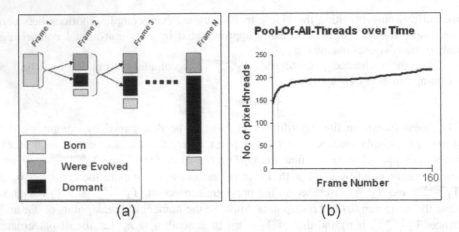

Fig. 4. The process of birth and evolution of *pixel-threads* across the frames to create the Generative Sketch-based Video (GSV) (a) The vertical bars represent the state of the *Pool-of-All-Pixel-Threads* Ψ_i for frame i; (b) For a video sequence of 160 frames, the number of *pixel-threads* in each frame is plotted as a function of the frame number (time)

In the next section, we present a method to evolve an entire generation of *pixel-threads* as a function of time. This results in the generation of a sketch-based representation of the original video sequence.

3.3 Evolution of a Pixel-Thread-Pool

Given a video sequence of N frames, and the current frame j, let Ψ be the pool of all the *pixel-threads* which have been *born* or *evolved* thus far in frames 1 through j -1. All the *pixel-threads* in Ψ may not be active, i.e., some may be dormant. The dormant *pixel-threads* still belong to Ψ, but represent *pixel-threads* which were not used to sketch a curve in the previous frame, j -1. The *pixel-threads* in Ψ belong to one of two subsets; $\Psi_{dormant}$ or Ψ_{active}. Clearly, $\Psi = \Psi_{dormant} \cup \Psi_{active}$.

For the current frame j, the *pixel-threads* corresponding to frame j are first determined using the techniques discussed in Section 3.1. These recently acquired *pixel-threads* corresponding to frame j are grouped together in *Pixel-thread-Pool* TP_j. Assume that TP_j has n_j *pixel-threads* $\{T^1_j, T^2_j \ldots T^{nj}_j\}$. Next, the correspondence between the *pixel-threads* in the set Ψ and the n_j *pixel-threads* in TP_j, is determined using the methods mentioned in Section 3.2. Note that per the terminology in Section 3.2, Ψ corresponds to TP_1 and TP_j corresponds to TP_2. Note that, during the correspondence determination procedure, the dormant *pixel-threads* in Ψ are also considered.

Let $TP_j^{wereEvolved}$ be the subset of the *pixel-threads* in TP_j which have evolved from Ψ. Let Ψ_{TPj} be the subset of *pixel-threads* in Ψ which evolve to a corresponding *pixel-thread* in $T P_j^{wereEvolved}$. It must be noted that, the correspondence between Ψ_{TPj} and $TP_j^{wereEvolved}$ is determined in a manner such that a *pixel-thread* T^i_j belonging to $TP_j^{wereEvolved}$ and the corresponding *pixel-thread* in Ψ_{TPj} from which it has evolved are both assigned the same index i. Now, *pixel-threads* in Ψ_{TPj} can be evolved to corresponding *pixel-threads* in $TP_j^{wereEvolved}$ via a set of motion models $\Theta_{\Psi_{TPj}}$

(equation (4)). Since the remaining *pixel-threads* in TP_j^i cannot be evolved from any existing *pixel-thread* in Ψ, these *pixel-threads* are considered to belong to the set TP_j^{born}; where $TP_j^{born} = TP_j - TP_j^{wereEvolved}$.

Next, the set Ψ is updated in the following manner:

(a) *Pixel-threads* in Ψ_{TPj} are *evolved* to corresponding *pixel-threads* in $TP_j^{wereEvolved}$, using motion model parameters given by $\Theta_{\Psi TPj}$. The new *pixel-threads* in TP_j^{born} are included in Ψ.

(b) The new set of active *pixel-threads* is given by $\Psi_{active} = \Psi \cap TP_j$. These *pixel-threads* are used to generate the sketch-based representation of the new video frame. Naturally, the *pixel-threads* in this updated set Ψ, that have no counterparts in TP_j, are deemed *dormant*; i.e.,

$$\Psi_{dormant} = \Psi - \Psi_{active}$$

The data corresponding to frame j required to sketch the jth video frame are given by the motion model parameters denoted by $\Theta_{\Psi TPj}$. The newly born *pixel-threads* are included in TP_j^{born}. Thus, the entire process of evolution of all the *pixel-threads* across all the N frames of the video can be effectively represented as a *Generative Sketch-based Video (GSV)*, given by

$$Generative\text{-}Sketch\text{-}based\text{-}Video = \{< \Theta_{\Psi TP1}, TP_1^{born}>,\ldots, < \Theta_{\Psi TPN}, TP_N^{born}>\} \quad (6)$$

A depiction of the process of evolution of the GSV is given in Fig 4(a). Fig 4(b) shows a plot of the total number of *pixel-threads* in Ψ as a function of time during the entire process of evolution of all the *pixel-threads* in the GSV representation of a sample video. The curve shows that, after the initial *pixel-threads* are created in the first frame, very few new *pixel-threads* are born thereafter. The initial *pixel-threads* are seen to be adequate to evolve and generate most of the entire GSV representation of the sample video.

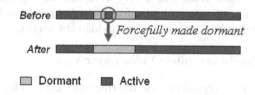

Fig. 5. Flicker removal using the history of activity or dormancy of a *pixel-thread*. The encircled portion corresponds to a brief period of activity for the *pixel-thread*. The *pixel-thread* is made dormant to remove the flickering effect caused by this brief activity.

3.4 Flicker Reduction

When the *pixel-threads* for each frame are rendered, flickering effects are observed. This is due to the fact that some pixel threads appear momentarily in a frame, only to become dormant in a series of successive frames. The resulting sudden appearance and disappearance of *pixel-threads* creates a flickering effect. The *All-Threads-Pool* Ψ contains the list of all the dormant *pixel-threads*. When a *pixel- thread*, which is dormant for some time, becomes active for a few frames, and then becomes dormant again, a flickering effect is observed. Thus, if such a *pixel-thread* is forced to be dormant instead of becoming active for a short time, the flickering effect is considerably reduced. The history of activity and dormancy is maintained for each *pixel-thread* in each frame, while the frame is being encoded. Once the entire video has been encoded, a second pass is made to determine, for each *pixel-thread*, the frames in which the *pixel-thread* should be made dormant, using the algorithm depicted in Fig 5.

4   Encoding the Texture - $V_{TEXTURE}$

In the previous section, we described in detail how a Generative Sketch-based Video (GSV) is obtained from the original video. The GSV is used as the *sketch* component within the proposed HLV representation. In this section, we describe how the second component of HLV, i.e., the *texture* component, is created. The texture of the video, given by the component $V_{TEXTURE}$, consists of three sub-components termed as V_{org}, V_{mid} and V_{base} where V_{org} is the original video which is deemed to be of the highest visual quality, V_{mid} is the video of intermediate visual quality, and V_{base} is the base-level video of the lowest visual quality. All the above video sub-components are encoded using the MPEG H.264 standard. In the interest of inter-operability, a deliberate effort has been made to incorporate within the proposed HLV encoding scheme, as many features of the MPEG H.264 standard as possible. In this section, we will discuss in detail the procedure for generation of each of the three layers.

4.1   Generating the Top-Most Video Layer V_{org}

V_{org} is the original video which is encoded efficiently using a state-of-the-art MPEG H.264 encoder that is available in the public domain [17]. A raw video encoded using the MPEG H.264 codec results in a very compact file representation. The MPEG H.264 codec uses inter-frame and intra-frame predictions to reduce significantly the spatial and temporal redundancy in the input video stream [8], [28].

4.2   Generating the Intermediate Video Layer V_{mid}

The video layer V_{mid} represents an intermediate-level video which has a more compact representation than the original video albeit at the cost of lower visual quality. It is a common observation that a lower-size video file leads to reduction in overall power consumption during the decoding process. The video layer V_{mid} is generated using a novel multi-resolution video encoding technique termed as *Features, Motion and Object-Enhanced Multi-Resolution* (FMOE-MR) video encoding [5]. The FMOE-MR video encoding scheme is based on the fundamental observation that applying a low pass filter in the image color space is equivalent to DCT coefficient truncation in the corresponding DCT (frequency) space [14]. The FMOE-MR video encoding scheme is a two step process as described in the following subsections.

Generating the FMO-Mask. Instances of Features, Motion and Objects (FMOs) are detected in the video sequence using state-of-the-art computer vision algorithms. A corresponding mask, termed as an *FMO-Mask*, is created to mark the regions corresponding to the presence of the FMOs. The mask contains floating point values between (and inclusive of) 0 and 1, where 0 represents a completely uninteresting region and 1 represents a region that is vital for visual and semantic understanding of the image. The *FMO-Mask* is essentially a combination of one or more of the following three individual masks; the *Feature-Mask (F-Mask)*, *Motion-Mask (M-Mask)* and the *Object-Mask (O-Mask)*.

Feature-Mask (F-Mask): The *Feature-Mask* captures the important low-level spatial features of the video frame. Edges are one of the most important low-level features of an image (or video frame); since human perception tends to first detect edges for the purpose of object recognition and general scene analysis. Edges can be detected automatically in a given image or video frame. There are many ways to perform edge detection. However, the majority of different methods may be grouped into two broad categories: gradient-based and Laplacian-based. The gradient-based methods detect the edges by seeking a maximum in the magnitude of the first derivative of the image intensity (or color) function. The Laplacian-based methods, on the other hand, search for zero crossings in the second derivative of the image intensity (or color) function to detect and localize the edges. In our current work we have used the Canny edge detector [4] which is a gradient-based method to detect and localize the edges in a video frame. Once the edges in the video frame are detected and localized using the Canny edge detector, an F-Mask is created by assigning a value of 1 to regions in and around the edges, and the value of 0 elsewhere. Note that the mask is essentially a weighting matrix, where each pixel may be assigned a value between (and inclusive of), 0 and 1.

Motion-Mask (M-Mask): The motion within a video sequence constitutes a very important visual phenomenon since human perception of a dynamic scene tends to follow the moving objects and note their activities. Therefore, in situations which demand reduction in quality of the video, the image regions in the video that are characterized by significant motion can be rendered at high resolution and the remainder of the video frame at low resolution. A *Motion-Mask (M-Mask)* is obtained by identifying the regions within the video frames that contain moving objects. This is essentially accomplished via a process of background subtraction [7], [18], [19], [23], [24]. Background subtraction is performed typically by first creating (or learning) a background model for a video sequence. The video frames are then compared with the background model to detect regions which are not part of background. These regions are classified as belonging to the dynamic foreground, i.e., containing moving objects. Background subtraction thus allows one to extract foreground objects which are moving relative to the camera, or had been moving until recently. We used the background models described in [23], [24] to extract the dynamic foreground (i.e., the moving objects) from the background.

Object-Mask (O-Mask): All the foreground objects in a video sequence may not be equally important from a visual or semantic perspective. For example, in a video sequence containing a news reader with a rotating background logo, the face of the news reader is more important than the moving logo which, in the current implementation, is also considered part of the foreground. In this case, the face is

deemed an object of interest amongst the various foreground regions. Face detection
and tracking is typically done by extracting feature points that characterize a face, and
tracking these feature points in the video sequence. A detailed description of the face
recognition and tracking algorithm is once again beyond the scope of this paper. In
our current implementation, faces in a video sequence are detected using the
algorithms described in [34], [35] and tracked using the algorithm described in [25] to
create an *O-Mask* based on human faces automatically.

Combining F, M, O masks to form a single FMO-Mask: The three masks are
superimposed to generate the final *Features-Motion-Object-Mask* (*FMO-Mask*). It is
not always required to generate all the masks; for example, for a surveillance
scenario, only the motion mask is required to capture the moving persons. Creation of
the mask also depends on the computational resources available; the F-Mask and M-
Mask can be typically generated in real time, and also combined in real time. When
the resources available at the coding end are significantly constrained, only the F-
Mask can be used since the edge detection procedure via the Canny edge detector is a
task of low computational complexity.

Multi-Resolution (MR) re-encoding. The original frame is re-encoded as a multi-
resolution (MR) representation, guided by the FMO-Mask such that regions
corresponding to mask values close to 1 are at higher resolution than regions
corresponding to mask values close to 0. The original video frame V_O is used to
render two video frames, V_H and V_L, such that V_H is a high resolution rendering and
V_L is a low resolution rendering of V_O. The video frames V_L and V_H are obtained by
convolving V_O with Gaussian filters characterized by the smoothing parameters σ_L
and σ_H respectively. Maintaining $\sigma_L > \sigma_H$ ensures that V_L is smoother than V_H, i.e., V_L
is a lower resolution rendering of V_O than V_H. If the FMO-Mask is represented as a
matrix W whose elements lie in the range [0, 1], then the MR frame V_{MR} is obtained
via a linear combination of the two frames V_H and V_L as follows:

$$V_{MR} = W \bullet V_H + (I - W) \bullet V_L$$

where I is a matrix all of whose elements are 1.

Fig. 6. Demonstration of the effect of Features, Motion and Object-enhanced Multi-Resolution
(FMOE-MR) video encoding. (a) The original frame; (b) The FMO mask frame; (c) The frame
re-rendered using FMOE-MR video encoding. Moving objects are rendered at high resolution
whereas the background is rendered at low resolution. The obtained PSNR is 24.94.

The values of σ_L and σ_H used to generate $V_{mid} = V_{MR}$ are selected empirically by the user. Empirical observations have revealed that $\sigma_L = 9$ or 11, and $\sigma_H = 3$, can be used, in most cases, to yield videos of reasonable visual quality with significantly smaller file sizes than the original video. Fig 6 presents an example of a video frame where the foreground moving object has been extracted using an FMO-mask. As is apparent, the regions where the moving objects are situated are rendered at higher resolution compared to the stationary regions comprising the background. Since an exhaustive treatment of the FMOE-MR video encoding scheme is beyond the scope of this paper, the interested reader is referred to [5] for further details. It must be noted that finally, V_{mid} too is encoded using the standard MPEG H.264 encoder, after preprocessing using FMOE-MR.

4.3 Generating the Base Video Layer V_{base}

The base video layer V_{base} is generated by first blurring each frame of the video using a Gaussian filter with smoothing parameter σ_{base} *prior* to MPEG H.264 encoding. Note that this is similar to the Gaussian smoothing procedure performed in the case of FMOE-MR video encoding. The primary difference is that, in the case of the base video layer generation procedure, the smoothing is performed uniformly over the entire video frame in contrast to FMOE-MR video encoding where the extent of smoothing can vary within a video frame based on the perceptual significance of the region under consideration. This results in further dramatic decrease in the file size upon MPEG H.264 encoding, albeit at the loss of video quality. Note that V_{base} is at of much lower visual quality than V_{mid} since object-based enhancement is not used. V_{base} essentially serves to provide approximate color information for the Generative Sketch-based Video (GSV) representation described previously.

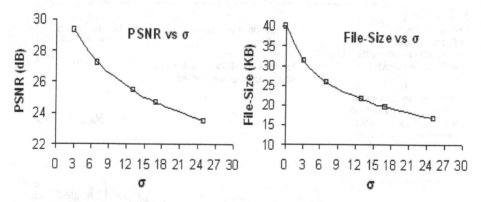

Fig. 7. The change in PSNR and file size of the video as a function of the Gaussian smoothing parameter σ_{base}

4.4 Assessment of Visual Quality of $V_{TEXTURE}$

The visual quality of each of the aforementioned video layers comprising $V_{TEXTURE}$ can be assessed in terms of PSNR values, as well as via subjective visual evaluation.

A quantitative evaluation of the average PSNR of a sample video with respect to the percentage decrease in video size is depicted in Fig 7. It is apparent that the video size can be decreased significantly by using a high value of σ_{base}, albeit with a loss in video quality. We have observed empirically that values of σ_{base} in the range [19, 25] can be used to generate the base video layer V_{base}, resulting in a very small file size albeit at the cost of low resolution and low visual quality. However, approximate color information is still retained in the video layer V_{base}, to the point that the visual quality of the resulting video improves significantly when the object outlines from the GSV representation are superimposed on the video layer V_{base}. Fig 8 depicts the overall video encoding procedure. The specific HLV is generated by overlaying the sketch component V_{SKETCH} over an appropriately chosen layer of $V_{TEXTURE}$ i.e., V_{org}, V_{mid} or V_{base} in our case. In the current implementation, the HLV encoding is done off-line. Consequently, the run times of the various procedures for generating the GSV and each of the texture layers V_{org}, V_{mid} and V_{base} are not very critical. Future work will focus on enabling real-time HLV encoding. In the next section, it is shown how different combinations of V_{SKETCH} and $V_{TEXTURE}$ result in distinct video states where each video state has a characteristic resource consumption profile.

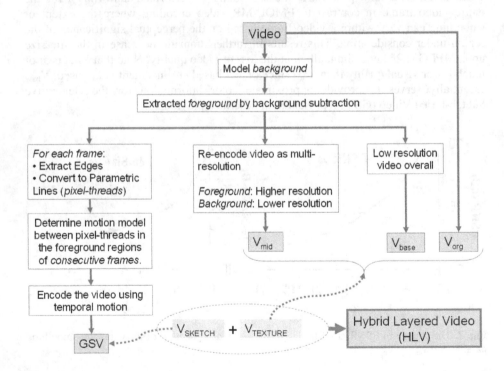

Fig. 8. The change in PSNR and file size of the video as a function of the Gaussian smoothing parameter σ_{base}

5 Encoding the Texture – V$_{\textbf{TEXTURE}}$

As mentioned in the previous sections, the two components, **V**$_{\textbf{TEXTURE}}$ and **V**$_{\textbf{SKETCH}}$, are obtained independently of each other. First, a suitable texture frame is extracted from the **V**$_{\textbf{TEXTURE}}$ component of the video by the video display controller. After this frame has been written to the frame buffer, the other component **V**$_{\textbf{SKETCH}}$, is used to superimpose the object outlines on the frame buffer containing **V**$_{\textbf{TEXTURE}}$. Note that both the aforementioned events are independent in terms of processing and are related only by order, i.e., **V**$_{\textbf{TEXTURE}}$ is rendered first, followed by the superimposition of **V**$_{\textbf{SKETCH}}$ on **V**$_{\textbf{TEXTURE}}$. An example frame obtained by superimposing **V**$_{\textbf{SKETCH}}$ on the V$_{\text{base}}$ subcomponent of **V**$_{\textbf{TEXTURE}}$ is shown in Fig 9.

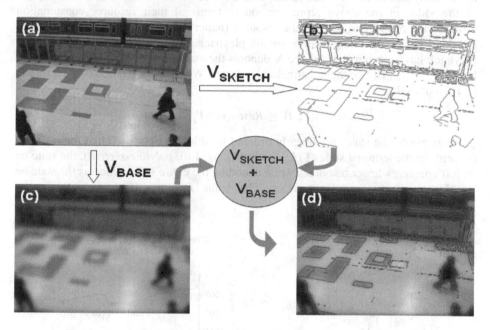

Fig. 9. An example demonstrating the combination of V$_{\text{SKETCH}}$ and V$_{\text{BASE}}$

Let us suppose that the **V**$_{\textbf{TEXTURE}}$ component has L levels of resolution. In the current implementation, $L = 4$ which includes the three layers V$_{\text{org}}$, V$_{\text{mid}}$ and V$_{\text{base}}$ in decreasing order of visual quality and level 0 which denotes complete absence of texture information. Let **V**$^{j}_{\textbf{TEXTURE}}$ $(0 \leq j \leq L\text{-}1)$ correspond to the MPEG H.264-based encoding of the video where **V**$^{0}_{\textbf{TEXTURE}}$ denotes the complete absence of texture information, **V**$^{1}_{\textbf{TEXTURE}}$ denotes the video layer of the least visual quality and resolution (with deliberately induced loss in visual quality to ensure a very low bitrate and small video file size), and **V**$^{L\text{-}1}_{\textbf{TEXTURE}}$ denotes the video layer of the highest visual quality and resolution (i.e., the original video encoded using the MPEG H.264 standard with no deliberately induced loss in visual quality). Let the state of the HLV-encoded video be depicted as:

$$\Gamma(\textit{texture-level, sketch-level}) = (V^{\text{texture-level}}_{\text{TEXTURE}}, V^{\text{sketch-level}}_{\text{SKETCH}}) \qquad (7)$$

such that $0 \leq \textit{texture-level} \leq L\text{-}1$, and $\textit{sketch-level} \in \{\textit{no-sketch, polyline-sketch,}$ $\textit{spline-sketch}\}$. The above state-based representation allows for various resolutions of texture with superimposition of sketch-based representations of varying degrees of complexity. Under the above formalism, $\Gamma(L, \textit{no-sketch})$ represents the original (i.e., best quality) video, and $\Gamma(0, \textit{polyline-sketch})$ represents the video that contains no texture, but only the object outlines represented by polylines (presumably, the lowest quality video).

Furthermore, the states in the above representation are ordered linearly such that a "higher" video state is deemed to consume more resources (battery power and bandwidth) than a "lower" video state. Thus, it is essential to order the different states of the video in the above representation in terms of their resource consumption profiles. Let $Resources(\mathbf{X}, t)$ be the resource (battery time, bandwidth, etc.) estimate provided by the operating system on the playback device t seconds after the video playback has been initiated, where $\mathbf{X}$ denotes the state of the video during playback. Let $\Gamma = \{\Gamma_1, ..., \Gamma_S\}$ be the S distinct video states. We define a relation $\leq_p$ such that $\Gamma_i \leq_p \Gamma_j$ implies that

$$Resources(\Gamma_i, t) \leq Resources(\Gamma_j, t), \quad t > 0 \qquad (8)$$

In other words, the states are linearly ordered from left to right such that for any state (except for the terminal states $\Gamma(L, \textit{no-sketch})$ and $\Gamma(0, \textit{polyline-sketch})$), the state on its left consumes fewer resources while decoding the entire video whereas the state on its right consumes more resources. The value $Resources(\Gamma_{\text{current-state}}, t)$ can be

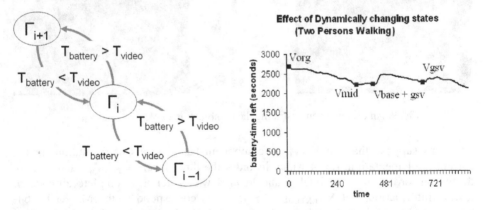

Fig. 10. (a) State diagram depicting the state transition rules based on available residual battery time. The current state transitions to a higher state if the available residual battery time (T_{battery}) is greater than the remaining running time of the video (T_{video}). Similarly, the current state transitions to a lower state if $T_{\text{battery}} < T_{\text{video}}$. (b) The effect of video state transitions on the residual battery time. The video playback starts with $\Gamma(V_{\text{org}}, \textit{no-sketch})$, changes state to $\Gamma(V_{\text{mid}}, \textit{no-sketch})$ then to $\Gamma(V_{\text{base}}, \textit{spline-fit})$ and finally to $\Gamma(0, \textit{spline-fit})$. It is apparent that there is an improvement in the residual battery time every time a lower state is chosen.

estimated using simple operating systems calls, which predict the remaining (residual) battery time or available bandwidth based on the current system load. Fig 10(a) depicts the state transition rules based on the available residual battery time. Fig 10(b) depicts the effect of the state transitions on the residual battery time on an IBM Thinkpad laptop PC with a 2 GHz Centrino Processor, 1 GByte RAM and 250 GByte hard disk running the video clip, a frame of which is depicted in Fig 6. It is clear that there is a significant improvement in the available residual battery time every time the system transitions to a state with lower resource consumption. Experimental results on a collection of 10 different video clips have shown that it is possible to save 45 minutes of battery capacity (energy) on average for the aforementioned IBM ThinkPad laptop PC by using the lowest layer of the proposed HLV representation, comprising only of the sketch component without any texture content, when compared to the reception and playback of the original video. Adding the approximated texture to the sketch component is observed to result in an average residual battery time savings of 30 minutes when compared to the reception and playback of the original video. The above experimental results serve to validate the claim that the proposed HLV representation does indeed result in different resource consumption estimates for distinct states in the aforementioned video state-based representation.

(a)   (b)   (c)   (d)

Fig. 11. (a) Video frame from Γ_6^{train} (b) Result of background subtraction on Γ_6^{train} (c) Video frame from Γ_3^{train} (d) Result of background subtraction on Γ_3^{train}

6 HLV for Mobile Internet-Based Multimedia Applications

Note that the various video layers (or video states) in the proposed HLV encoding scheme are generated by altering the quality of the resulting encoded video. Since a video state in the proposed HLV encoding scheme is only an approximation to the original MPEG-encoded video, it raises a natural question, i.e., is the approximation good enough? Subjective evidence gathered from various human subjects has revealed that all the objects discernable in the MPEG-encoded video are also discernable in the HLV-encoded video. However, in the interest of objectivity, we have evaluated the proposed HLV encoding scheme in the context of some important multimedia applications in a resource-constrained mobile Internet environment, i.e., background subtraction, face detection, face tracking and face recognition, using quantitative performance metrics. Our current experiments are limited to the measurement of resource (battery power and bandwidth) consumption on the mobile end-user device on which the video is decoded and rendered. All experiments were performed on the aforementioned IBM Thinkpad laptop PC running in battery mode. In our future work, we intend to include other types of mobile devices such as PDAs, iPhones and pocket-PCs in our experiments.

6.1 Background Subtraction

As an objective comparison of HLV-encoded video quality, we used the videos corresponding to video states of the *Train Station* video, $\Gamma_3^{train} = \Gamma(V_{base}, polyline-sketch)$, and $\Gamma_6^{train} = \Gamma(V_{org}, null)$, to perform background subtraction. Note that $\Gamma_6^{train} = (V_{org}, null)$, in our case, corresponds to the video with the highest visual quality. Background subtraction is the process of first estimating the background of the

Fig. 12. Video 1: Select frames from the face tracking experiment

Fig. 13. Video 2: Select frames from the face tracking experiment

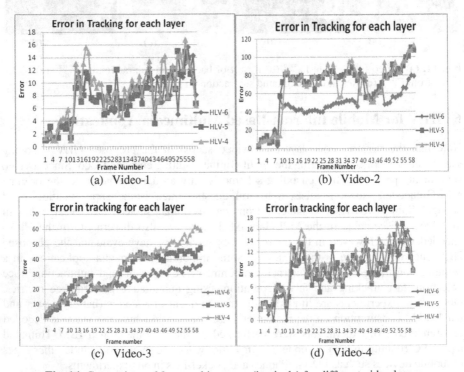

(a) Video-1   (b) Video-2

(c) Video-3   (d) Video-4

Fig. 14. Comparison of face tracking error (in pixels) for different video layers

dynamic scene, where the background is deemed to comprise of those regions within the video frames which do not move relative to the camera. The background, thus determined, is subtracted from each frame to extract the foreground, or moving regions within the video frames [23], [24]. We hypothesize that the video states Γ_3^{train} and Γ_6^{train} yield videos of comparable quality if both videos result in similar foreground regions upon background subtraction.

Fig 11 shows the resulting foreground masks after background subtraction has been performed on a Γ_6^{train} video frame and the corresponding Γ_3^{train} frame. As is evident from Fig 11, both videos were observed to yield similar foreground masks. We further computed the percentage overlap between the foreground masks generated from the two videos. The mask generated from the Γ_3^{train} video frame was observed to have an 85% overlap with the mask generated from the original video, Γ_6^{train}. Note that it is not possible to use standard metrics such as the Peak Signal-to-Noise Ratio (PSNR) to measure the quality of HLV-encoded video, since the PSNR measure treats the graphics overlay of outlines, i.e., V_{SKETCH}, as noise. The PSNR-based quality measure is limited to $V_{TEXTURE}$ as shown in Fig 7. Quality assessment of the proposed HLV encoding scheme would involve an extensive survey of a significant user population. In such a survey, the users would be required to provide qualitative feedback on their viewing experience for the different HLV states and compare their viewing experience with HLV-encoded video to that with conventional MPEG-encoded video. Such a user survey is beyond the scope of the current work and will be pursued in our future research.

6.2 Face Detection, Face Recognition and Face Tracking

The proposed HLV encoding scheme was also evaluated in the context of face detection, face tracking and face recognition. Four videos of different people were taken and three different HLV layers were used, i.e., $\Gamma_6^{face} = (V_{org}, null)$, $\Gamma_5^{face} = (V_{mid}, spline-sketch)$ and $\Gamma_4^{face} = (V_{mid}, spline-sketch)$. Face detection was performed using a combination of color-based skin detection [35] and Haar features used in the AdaBoost face detection technique [34]. The color-based skin detection algorithm [35] was used to provide regions of interest for the AdaBoost face detection algorithm. In the AdaBoost face detection algorithm, a 25-layer cascade of boosted classifiers was trained to detect multi-view faces. A set of sample face images and a set of non-face (i.e., background) images were used for training. Each sample image was cropped and scaled to a fixed resolution. A set of multi-view face images was collected from video sequences under different conditions of ambient scene illumination, surface reflection, facial pose, facial expression and background composition in order to make the face detection procedure more robust in different scenarios and under different scene conditions. Another set of non-face sample images was collected from video sequences containing no faces.

The detected faces were tracked using the kernel-based mean-shift tracker [9]. The mean-shift tracker is an iterative region localization procedure based on the maximization of a similarity measure, such as one based on the Bhattacharyya distance, between the color distributions or color histograms of the tracked regions (containing human faces) in successive video frames. Fig 12 and Fig 13 show select frames from two of the four sample video clips (Video 1 and Video 2) used in the face tracking experiment. Fig 14 (a)-(d) show the face tracking error for the different HLV layers on all four sample video clips where the layers Γ_6^{face}, Γ_5^{face} and Γ_4^{face} are denoted by HLV-6, HLV-5 and HLV-4 respectively. The tracking error in the case of Γ_6^{face} (HLV-6) is indicative of the error introduced by the face detection and tracking algorithms and the basic MPEG encoding scheme. As can be seen from the plots for Γ_5^{face} (HLV-5) and Γ_4^{face} (HLV-4) very little additional error is introduced by the proposed HLV encoding scheme. As can be seen from Fig 14(a) and Fig 14(d), the tracking error values are almost similar for all three HLV layers under consideration. Note that the tracking error for each video frame is determined by computing the Euclidean distance between the centroid of the face region output by the face detection algorithm (in the first frame) or face tracking algorithm (in subsequent frames) and the centroid of the corresponding face region that is manually delineated. Video 1 and Video 4 are examples of video clips containing a single face whereas Video 2 and Video 3 contain multiple faces with significant scene clutter. On account of occlusion and scene clutter, the tracking error in Video 2 and Video 3 was observed to be significantly higher than that in Video 1 and Video 4.

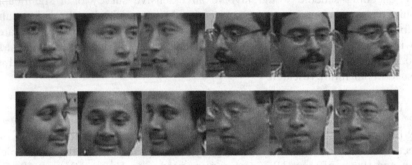

Fig. 15. Examples of face images used for training the *Eigenfaces* algorithm

The HLV scheme was also evaluated in the context of face recognition performed using the well known *Eigenfaces* approach [33]. The eigenfaces are essentially eigenvectors that are derived from the covariance matrix of the probability distribution of the high-dimensional vector space of human faces (with known identity) that are stored in a database. The eigenfaces constitute a basis set of "standardized faces", derived from the training set of human faces in the database via principal component analysis. The training set of human faces is used to estimate the mean vector and covariance matrix of the probability distribution of the high-dimensional vector space of human faces stored in the database under varying conditions of illumination, viewpoint and pose. Each human face in the database can be expressed as a unique linear combination of these eigenfaces. An unknown face is also expressed as a linear combination of these eigenfaces. The eigenface coefficients (eigenvalues) of the unknown face are compared with those of each of the faces in the database by computing the Euclidean distance in the vector space spanned by the eigenfaces. The unknown face is recognized via the identity of the closest face in the database. The *Eigenfaces* approach is essentially an appearance-based face recognition method.

In our implementation, the face recognition system used a database of grayscale images scaled to 64 × 64 pixels with frontal profiles of 4 persons and 10 variations in pose, lighting and background conditions for each person. Fig 15 shows some of the faces used for training the *Eigenfaces* algorithm. The graph in Fig 16 compares accuracy of face recognition for each of the HLV layers under consideration, i.e., HLV-4, HLV-5 and HLV-6. The recognition accuracy was tested for varying ratios of the number of training images to the number of test images. All the training images were derived from the video layer HLV-6 whereas face recognition was performed on video frames derived from all the three HLV layers under consideration. As can be seen, the recognition accuracy decreases only slightly in the case of video layer HLV-4 whereas there is no major difference in face recognition accuracy between the video layers HLV-5 and HLV-6.

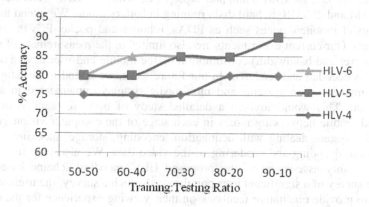

Fig. 16. Comparison of % accuracy in face recognition for different video layers

The above experimental results show that the proposed HLV encoding scheme is well suited for mobile Internet-based multimedia applications that are typically constrained by available resources, in particular, the available bandwidth and battery capacity of the mobile client device.

7 Conclusion

The increasing deployment of broadband networks and simultaneous proliferation of low-cost video capturing and multimedia-enabled mobile devices has triggered a new wave of mobile Internet-based multimedia applications. Applications such as Internet-based video on demand, community-based video sharing, and multimedia web services are no longer mere research ideas, but have resulted in several commercial products. However, mobile networked environments are typically resource constrained in terms of the available bandwidth and battery capacity on mobile devices. Multimedia applications that typically entail analysis, transmission, storage and rendering of video data are resource-intensive. Since the available bandwidth in the mobile Internet is constantly changing and the battery life of a mobile video capturing and rendering device decreases with time, it is desirable to have a video representation scheme that adapts dynamically to the available resources. To this end we have proposed and implemented a *Hybrid Layered Video* (HLV) encoding scheme, which comprises of content-aware, multi-layer encoding of texture and a generative sketch-based representation of the object outlines. Different combinations of the texture- and sketch-based representations result in distinct video states, each with a characteristic bandwidth and power consumption profile.

The proposed HLV encoding scheme is shown to be effective for mobile Internet-based multimedia applications such as background subtraction, face detection, face tracking and face recognition on resource-constrained mobile devices. Future work will consider more advanced computer vision-based applications in a mobile Internet environment such as intelligent video surveillance, distributed gaming and human activity understanding. Our current experiments are limited to a single, potentially mobile device, i.e., an IBM Thinkpad laptop PC, with a 2 GHz Centrino CPU, 1 GByte RAM and 250 GByte hard disk, running in battery mode. We intend to include other types of mobile devices such as PDAs, iPhones and pocket-PCs in our future experiments. Our current experiments are also limited to the measurement of resource (battery power and bandwidth) consumption on the mobile end-user device on which the video is decoded and rendered. In our future work, we intend to investigate end-to-end computer vision systems and multimedia systems implemented in a mobile environment. This would involve a detailed study of both, resource consumption issues and mobile networking issues in each stage of the computer vision system or multimedia system dealing with acquisition, encoding, storage, indexing, retrieval, transmission, decoding and rendering of the video data. We also plan to perform a thorough quality assessment of the proposed HLV encoding scheme based on an extensive survey of a significant user population. In such a survey, the users would be required to provide qualitative feedback on their viewing experience for the different HLV states and compare their viewing experience with HLV-encoded video to that with conventional MPEG-encoded video. The aforementioned survey would serve to ascertain the superiority of the proposed HLV encoding scheme to the conventional MPEG encoding standard. In the current implementation, the HLV encoding is done off-line. Consequently, the run times of the various procedures for generating the GSV and each of the texture layers V_{org}, V_{mid} and V_{base} are not very critical. Future work will focus on enabling real-time HLV encoding.

References

1. Atallah, M.J.: Linear time algorithm for the Hausdorff distance between convex polygons. Information Processing Letters 17(4), 207–209 (1983)
2. Besl, P.J., Jain, R.: Segmentation through variable-order surface fitting. IEEE Trans. Pattern Analysis and Machine Intelligence 10(2), 167–192 (1988)
3. Bouguet, J.Y.: Pyramisdal Implementation of the Lucas Kanade Feature Tracker, Intel Corporation, Microprocessor Research Labs; included in the distribution of OpenCV
4. Canny, J.: A computational approach to edge detection. IEEE Trans. Pattern Analysis and Machine Intelligence 8(6), 679–698 (1986)

5. Chattopadhyay, S., Luo, X., Bhandarkar, S.M., Li, K.: FMOE-MR: content-driven multi-resolution MPEG-4 fine-grained scalable layered video encoding. In: Proc. ACM Multimedia Computing and Networking Conference (ACM MMCN 2007), San Jose, CA, January 2007, pp. 650404.1–11 (2007)
6. Chattopadhyay, S., Bhandarkar, S.M., Li, K.: Ligne-Claire video encoding for power constrained mobile environments. In: Proc. ACM Multimedia, Augsburg, Germany, September 2007, pp. 1036–1045 (2007)
7. Cheung, S.-C., Kamath, C.: Robust background subtraction with foreground validation for urban traffic video. EURASIP Jour. Applied Signal Processing 14, 1–11 (2005); UCRL-JRNL-201916
8. Coding of Audio-Visual Objects, Part-2 Visual, Amendment 4: Streaming Video Profile, ISO/IEC 14 496-2/FPDAM4 (July 2000)
9. Comaniciu, D., Ramesh, V., Meer, P.: Kernel-based object tracking. IEEE Trans. Pattern Analysis and Machine Intelligence 25(5), 564–577 (2003)
10. Cornea, R., Nicolau, A., Dutt, N.: Software annotations for power optimization on mobile devices. In: Proc. Conf. Design, Automation and Test in Europe, Munich, Germany, March 2006, pp. 684–689 (2006)
11. Cucchiara, R., Grana, C., Prati, A., Vezzani, R.: Computer vision techniques for PDA accessibility of in-house video surveillance. In: Proc. ACM SIGMM International Workshop on Video Surveillance, Berkeley, CA, November 2003, pp. 87–97 (2003)
12. Dai, M., Loguinov, D.: Analysis and modeling of MPEG-4 and H.264 multi-layer video traffic. In: Proc. IEEE INFOCOM, Miami, FL, March 2005, pp. 2257–2267 (2005)
13. Davies, E.: Machine Vision: Theory, Algorithms and Practicalities, pp. 42–44. Academic Press, San Diego (1990)
14. Geusebroek, J.-M., Smeulders, A.W.M., Van de Weijer, J.: Fast anisotropic Gauss filtering. IEEE Trans. Image Processing 12(8), 938–943 (2003)
15. Hakeem, A., Shafique, K., Shah, M.: An object-based video coding framework for video sequences obtained from static cameras. In: Proc. ACM Multimedia, Singapore, November 2005, pp. 608–617 (2005)
16. Hennessy, J.L., Patterson, D.A.: Computer Architecture - A Quantitative Approach, 4th edn. Appendix D. Morgan Kaufmann, San Francisco (2007)
17. http://www.squared5.com
18. Ivanov, Y., Bobick, A., Liu, J.: Fast lighting independent background subtraction. International Journal of Computer Vision 37(2), 199–207 (2000)
19. Javed, O., Shafique, K., Shah, M.: A hierarchical approach to robust background subtraction using color and gradient information. In: Proc. IEEE Workshop on Motion and Video Computing, Orlando, FL, December 2002, pp. 22–27 (2002)
20. Khan, S., Shah, M.: Object based segmentation of video using color motion and spatial information. In: Proc. IEEE Conf. Computer Vision and Pattern Recognition, Kauai Island, HI, December 2001, vol. 2, pp. 746–751 (2001)
21. Ku, C.-W., Chen, L.-G., Chiu, Y.-M.: A very low bit-rate video coding system based on optical flow and region segmentation algorithms. In: Proc. SPIE Conf. Visual Communication and Image Processing, Taipei, Taiwan, May 1995, vol. 3, pp. 1318–1327 (1995)
22. Liang, C., Mohapatra, S., Zarki, M.E., Dutt, N., Venkatasubramanian, N.: A backlight optimization scheme for video playback on mobile devices. In: Proc. Consumer Communications and Networking Conference (CCNC 2006), January 2006, vol. 3(2), pp. 833–837 (2006)

23. Luo, X., Bhandarkar, S.M.: Robust background updating for real-time surveillance and monitoring. In: Proc. Intl. Conf. Image Analysis and Recognition, Toronto, Canada, September, 2005, pp. 1226–1233 (2005)
24. Luo, X., Bhandarkar, S.M., Hua, W., Gu, H., Guo, C.: Nonparametric background modeling using the CONDENSATION algorithm. In: Proc. IEEE Intl. Conf. Advanced Video and Signal-based Surveillance (AVSS 2006), Sydney, Australia, November 2006, pp. 13–18 (2006)
25. Luo, X., Bhandarkar, S.M.: Tracking of multiple objects using optical flow-based multiscale elastic matching. In: Vidal, R., Heyden, A., Ma, Y. (eds.) WDV 2005/2006. LNCS, vol. 4358, pp. 203–217. Springer, Heidelberg (2007)
26. Mohapatra, S., Cornea, R., Dutt, N., Nicolau, A., Venkatasubramanian, N.: Integrated power management for video streaming to mobile handheld devices. In: Proc. ACM Multimedia, Berkeley, CA, November 2003, pp. 582–591 (2003)
27. Ni, P., Isovic, D., Fohler, G.: User-friendly H.264/AVC for remote browsing. In: Proc. ACM Multimedia, Santa Barbara, CA, October 2006, pp. 643–646 (2006)
28. Richardson, I.E.G.: H.264 and MPEG-4 Video Compression: Video Coding for Next Generation Multimedia. Wiley, New York (2004)
29. Rosin, P.L., West, G.A.W.: Non-parametric segmentation of curves into various representations. IEEE Trans. Pattern Analysis and Machine Intelligence 17(12), 1140–1153 (1995)
30. Salembier, P., Marques, F., Pardas, M., Morros, J.R., Corset, I., Jeannin, S., Bouchard, L., Meyer, F., Marcotegui, B.: Segmentation-based video coding system allowing the manipulation of objects. IEEE Trans. Circuits and Systems for Video Technology 7(1), 60–74 (1997)
31. Salomon, D.: Data Compression: The Complete Reference. Springer, Berlin (2004)
32. Sikora, T.: Trends and perspectives in image and video coding. Proc. IEEE 93(1), 6–17 (2005)
33. Turk, M., Pentland, A.: Eigenfaces for recognition. Jour. Cognitive Neurosicence 3(1), 71–86 (1991)
34. Viola, P., Jones, M.: Rapid Object Detection using a Boosted Cascade of Simple Features. In: Proc. IEEE Conf. Computer Vision and Pattern Recognition, Kauai Island, HI, December 2001, vol. 1, pp. 511–518 (2001)
35. Zarit, B.D., Super, B.J., Quek, F.K.H.: Comparison of five color models in skin pixel classification. In: Intl. Workshop on Recognition, Analysis, and Tracking of Faces and Gestures in Real-Time Systems, Washington, DC, September 1999, pp. 58–63 (1999)

A Recommender Handoff Framework for a Mobile Device

Chuan Zhu[1], Matthew Ma[2], Chunguang Tan[3], Guiran Chang[3], Jingbo Zhu[3], and Qiaozhe An[3]

[1] Hohai University, China
zhu.ca@hotmail.com
[2] Scientific Works, USA
mattma@ieee.org
[3] Northeastern University, China
tcg1978@gmail.com, chang@neu.edu.cn,
zhujingbo@mail.neu.edu.cn, anqiaozhe@163.com

Abstract. Due to the proliferation of mobile phone market and the emergence of mobile TV standard, there is a need for providing consumers with electronic programming guide (EPG) on a mobile device. To reduce the information load on a mobile device, an EPG recommender framework is designed. The framework provides a flexible architecture to adapt to different environments whereas the recommender core can work stand-alone on a mobile device or switch to a hybrid mode utilizing the computing resource on a home network. In our prototype, a hybrid framework with automatic recommender handoff is built for DVB-H environment and simulated on latest Android mobile platform.

Keywords: Electronic programming guide, EPG, ESG, IPG, DVB-H, IPTV, Recommender, EPG proxy, Android.

1 Introduction

With the advancement of mobile and broadcasting technologies, mobile TV standard and products are emerging in the market. This makes traditional TV and data casting capabilities conveniently available on a mobile handset based on digital technologies. Among the competing standards DVB-H [1][2][3] is taking its presence in the mobile TV market. As the number of channels available on the existing and future TV broadcasting network increases, it becomes more challenging to deal with the overwhelmingly expanding amount of information provided by the electronic programming guide (EPG). Therefore, an EPG recommender system becomes eminent in meeting the needs of delivering a prioritized listing of programs.

Additionally, with the convergence of TV broadcasting for mobile device and cable network at home, the EPG recommender system has to adapt itself to a mobile environment. In one scenario, a user sharing the same TV with a family member may wish to use his/her mobile device as a secondary remote control and display. This secondary display can also be used for viewing supplementary information such as EPG, receiving coupons, and at the same time accessing viewing history and habit of

X. Jiang, M.Y. Ma, and C.W. Chen (Eds.): WMMP 2008, LNCS 5960, pp. 137–153, 2010.

oneself or children in the family [4][5]. In another scenario, when a mobile phone is equipped with TV broadcasting (Media-Flo or DVB-H), a user may watch TV from either mobile phone or home TV at different time of the day, and select a program of his or her preference in any environment, which requires the recommender system to work seamlessly and transparently across two environments.

In related literatures, some work have been described on TV recommendation schemes [6][7][8][9][10][11][12][13] using various technologies such as ontology, filtering, user profiling, relevance feedback etc.. Other work illustrated network platforms utilizing Internet [14][15], WAP/SOAP [16][17] and OSGi/SIP [10][18]. In comparison to prior approaches, our work focuses on providing a generic framework that is flexible and independent of networking protocols, broadcasting sources, and recommendation algorithms. It is also a precedent attempt as we are aware to simulate in a DVB-H environment.

In this chapter a hybrid EPG recommender framework previously proposed [19] is being further explored to support the EPG recommendation service on a mobile device anytime anywhere. The hybrid recommender framework is simulated in a DVB-H environment for automatic handoff of EPG service when a mobile device is roaming between outside and inside a home. The simulation results have shown that the proposed EPG recommender handoff framework is feasible.

The remainder of this chapter is organized as follows. Section 2 describes our proposed EPG recommender system and the mobile hand-off system architecture. Section 3 discusses implementation issues. Section 4 illustrates our prototype followed by conclusion in Section 5.

2 EPG Recommender Framework

2.1 An EPG Hybrid Framework

The so-called hybrid framework is intended to work in both mobile and home network modes. It is designed to support the following configurations: (1) A stand-alone mobile EPG application, in which an EPG service middleware is running on the mobile device without requiring any external data sources. This enables users to enjoy the EPG recommendation service anytime anywhere. (2) A client/server EPG application, for which a server (e.g. home server) provides an EPG recommender service on a more powerful device; and the mobile device acts as a browser in a slave mode displaying the recommended TV programs. This enables users to utilize a more powerful server to run a full scale complex recommender for his or her enjoyment of a better recommendation service. In the server mode, a variety of wireless protocols can be utilized by the mobile device to communicate with the server such as Bluetooth, Wi-Fi etc.

The hybrid EPG framework is shown in Fig. 1. As can be seen, the two blocks of mobile and server are clearly shown to support the above two scenarios. The detailed considerations of this framework are further described as below.

EPG Source. The proposed framework assumes that mobile device and home server have two independent EPG sources, and there are no EPG database transfers between these two devices. This would be a valid assumption for 1) Data casting via mobile network can easily upload an EPG database to the mobile device. Alternatively, the recent DVB-H makes it possible for a mobile phone to directly receive broadcasting TV and data stream in terrestrial domain. 2) As illustrated in Fig. 2, as the emerging DVB-H (or similar services) is taking maximum advantage of the existing Internet (e.g. Internet Protocol Data Casting, IPDC), we believe, in the near future, EPG from the same provider will be available on both mobile handset (e.g. DVB-H) and home (e.g. Internet, IPTV) simultaneously through broadband channel. This eliminates the need to transfer EPG from mobile to server or vice versa. Further, from commercial perspective, this assumption may also hold true with the increasing discussion and deployment of pioneering commercial enterprise, innovative experimentation and open-source community projects [20].

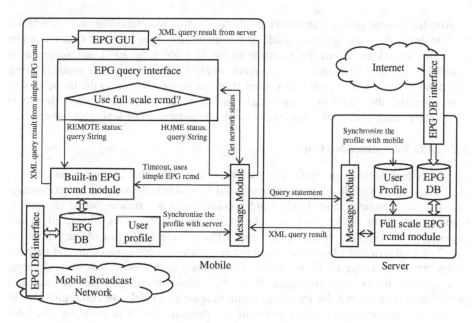

Fig. 1. A hybrid EPG recommender and its network architecture

The only EPG transfer that is allowed is the recommended EPG listing from home server to the mobile device. This reduces the complexity of the design and network latency. A crucial implementation issue around this would be a single user profile that fits into two forms of EPG: one from traditional broadcast (e.g. cable) and one from Mobile TV channels. This topic will be taken up later in Section 3.

EPG Recommender Middleware and Hand-off. A great benefit of our proposed EPG framework is its versatile configuration for adapting to different scenarios: a small foot-print integrating the simple built-in EPG recommender or a full scale EPG, all using the same application programming interface (API) [19]. The majority of functions in the EPG middleware are designed to share between mobile and server, and the EPG recommendation core can be implemented to adapt to two configurations, which will be discussed in the next sub-section.

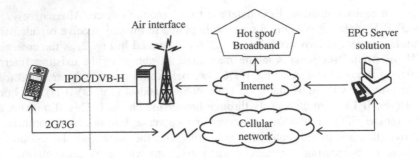

Fig. 2. Convergence of EPG to DVB-H and broadband

Another benefit of the framework is the handoff of recommendation task when a mobile device is roaming from outside of a home to inside the home or vice versa. For example, when user moves from outside to inside a home the EPG recommendation middleware works from mobile stand-alone mode to the client/server mode. During this handoff, only a user profile is copied from the mobile device to the home server. Consequently, the recommendation task that is running on the mobile device is completely taken over by the home server, which contains a full scale recommender.

2.2 EPG Recommender Core

As described above, an EPG recommender core can be implemented in two configurations: a simple recommender adapted to a mobile device and low computing resource environment; and a full scale recommender for running on a computing resource abundant device such as a home or media server. As illustrated in our previous work [9][19], an exemplary core EPG recommender module, as shown in Fig. 3, uses a series of filters to enhance the accuracy of recommendation and narrows down the search range of TV programs. Five filters: time, station, category, domain, and content filters, are implemented in the recommender module. A user can predefine a filter setting, for example, a date ranging two weeks, or certain number of channels. Category refers to the genre of the program and it is normally available from EPG data.

Domain information more closely reflects a user's domain of interest. It may cover programs across multiple genres. For example, "sports" domain includes any programs related to sports such as movie, news, game, etc. Domain information is normally not directly available from the content provider. It can be obtained via text classification of EPG content with corpus training [9].

The content filter is designed to recommend programs based on EPG content. It is more comprehensive as the EPG content comprises of all information in an EPG data such as stations, program titles, program descriptions, time interval, and actors. The content based filtering is advantageous over simple probability based filtering such that it does not require users to pre-select filtering criteria. The EPG recommendation core as illustrated in Fig. 3 can be configured to run on either mobile environment or server platform. For example, the simple recommender without content filter requires relative little computing resources and can be implemented fully in a mobile device. A full recommender with all filters including content filter is more viable to be deployed on a server environment with more computing resource.

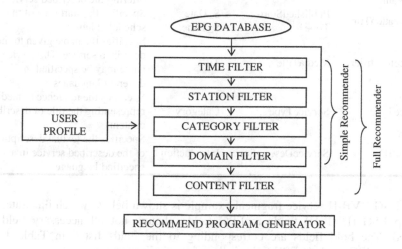

Fig. 3. A multi-engine EPG recommendation core

The proposed recommender framework is independent of recommendation engine thus the design of recommendation engine is outside the scope of this chapter. For reader's reference, we use a maximum entropy based classifier for both simple and full recommenders, and the accuracy of the full recommender was reported 75-80% in our previous work [19].

3 Implementation Issues

This section discusses any implementation issues relating to the network architecture and hand off operation, as well as various components of the recommender framework.

3.1 EPG Database

To be able to adapt to various EPG sources and isolate the dependence of EPG recommender core engine on different protocols, we have constructed a core EPG database (DB) that is simplified and suitable to our recommender. Any protocol specific EPG interface module (e.g. DVB-H, IPTV) needs to implement its own EPG parser and EPG proxy that map proper fields from electronic programming source to our core EPG database. In our reference implementation (prototype), the mapping from IPG (IPTV) and ESG (DVB-H) to our own EPG is conceptually illustrated in Table 1.

Table 1. EPG proxy field mapping and semantics table

IPG (IPTV)	ESG (DVB-H)	EPG	Semantics
IPTVExporter-HostName	ServiceName	Station	Specifies the service provider offering the described service.
RecCreationTime	PublishedStart-Time	S_Date S_Time	Specifies the start time of the scheduled item.
serviceIdentifier	MediaTitle	Title	Specifies the name given to the described service. The service name may be specified in different languages.
serviceType	ServiceType	Category	Specifies the reference to media representing the title of described content.
contentID	ServiceDescription	Introduction	Specifies the textual description of the described service in a specified language.

An ESG DVB-H service fragment example is shown below, which illustrates how to map ESG (DVB-H) fields to EPG fields, though not all necessary fields are covered. The bold fields are corresponding to the fields listed in Table 1. The mapping of IPG fields to EPG fields is done in a similar fashion.

ESG DVB-H service fragment example

```
<Service xmlns="urn:dvb:hhu:esg:2009"
      ServiceID="hhu.edu.cn/1013" clearToAir="1">
<ServiceName xml:lang="en">Example Service Name
      </ServiceName>
......
<ServiceDescription xml:lang="eng">Example description
      </ServiceDescription>
......
<ServiceType
      href="urn:dvb:hhu:esg:cs:ServiceTypeCS:1.2">
      <tva:Name xmlns:tva="urn:tva:metadata:2009">
            radio</tva:Name>
</ServiceType>
<ParentalGuidance>
      <mpeg7:MinimumAge
            xmlns:mpeg7="urn:mpeg:mpeg7:schema:2008">5
      </mpeg7:MinimumAge>
</ParentalGuidance>
<ServiceProvider>
      <ProviderName xml:lang="en">
      Example Content Provider Name </ProviderName>
      ......
</ServiceProvider>
      ......
</Service>
```

One implementation issue associated with EPG DB Interface is the synchronization of EPG downloaded from mobile to that downloaded from the server. We believe that this issue is vendor specific thus outside the scope of this chapter. Secondly, on the mobile device, the EPG download only practically occurs when the mobile is standing by and when no other critical applications (e.g. making a phone call) are running. Thirdly, EPG DB only keeps the necessary fields used for recommendation, plus a pointer for each item that corresponds to the program in original ESG, with which the mobile could address the content server and fetch broadcasting data.

3.2 User Profile

A user profile is used for recording user preferences to be used for the recommender engine. It records user's viewing history, for example, which specific program the user indicated he/she likes. In our prototype, a table in EPG DB with *hitTimes* and *like* fields does this job. *hitTimes* field is responsible for recording how many times the user clicks on an EPG item, while the *like* field is responsible for recording whether the user liked this program when he or she was browsing EPG list or watching the program. This is simply realized in prompting user to click "Like" or "Dislike" button when he or she is viewing the detailed synopsis of a particular TV program (as shown in Fig. 5).

Gadanho et. al have studied the impact of various implicit feedback strategies in a TV recommendation framework [21] and have shown little difference on the performance of these different strategies. Our "Like" and "Dislike" buttons seem to require an extra step for the user, however, in real deployment, after user views a particular program, he or she will either decide to watch the program (which shows a favor of "like") or skip to another program ("dislike"). Therefore, this feedback mechanism can be easily deployed into a completely implicit feedback in a real application without affecting the performance of recommendation. This implicit feedback is used to re-train the content based classifier, which accuracy increases over time.

From design perspective, a user profile is always stored on EPG server and will be used together with the full recommender core. When a mobile device is running EPG recommendation service in a client/server environment, the user profile on the server is used and updated. Due to the resource limitation on a mobile device, a simplified user profile storing partial data that is to be used for built-in simple recommender core, is residing on the mobile device and updated as well. Whenever the user profile on the server is updated, the simplified version is generated and synchronized with the mobile device. Therefore, the mobile device always keeps the latest user profile from the last connection to the server.

When a mobile device is not on the home network, this simple user profile is used for the built-in recommender and updated to record user's preferences on program listing. When the mobile device is connected to home network again, its simple user profile is synchronized to the server and will update the server with the most recent user profile.

Considering the link stability of wireless network, every time a mobile device tries to connect to the server, a successful connection can not be guaranteed. Frequent attempts to re-connect will cost battery power to the mobile device. So when and how to synchronize user profile is one of the key problems. In our prototype, to minimize consumption of power on a mobile device, synchronization of user profile only happens when a mobile device is using the EPG server. When a mobile device is requiring an EPG list from the EPG server, it sends not only the recommendation request, but also the simplified user profile in XML format. When the recommended list is received, a re-synchronized user profile is received from the server as well.

3.3 Handoff Operation

The hybrid EPG recommender architecture has two types of recommenders: the built-in simple recommender residing on a mobile device, and the full scale recommender residing on the home server. In order to properly use these two types of recommenders seamlessly, the working modes of a mobile device has three values: HOME/REMOTE/OFFLINE to indicate if the mobile device is inside home (server is available), outside home (server is not available), or no EPG service is activated, respectively.

In addition to monitoring the working mode of a mobile device, the status of network link between the mobile and the server is also monitored in order to guarantee the auto hand-off of EPG service. For example, when the EPG recommender is working in a server mode and the network is disconnected for some reason, the recommender needs to switch to the mobile mode right away such that the latency of EPG recommender service is not noticeable to the user.

The status of network link between mobile and server has three values - NORMAL, LINK_UNSTABLE, or SRV_UNAVAILABLE as described below:

1. NORMAL: When an EPG query is sent to EPG server, a complete XML format response is received by the mobile.
2. LINK_UNSTABLE: When an EPG query is sent to EPG server, an incomplete XML format response is received by the mobile. This can be caused by either unstable network or the length of the message exceeding the buffer size (set in UDP).
3. SRV_UNAVAILABLE: The EPG server is unreachable to the mobile. This can be caused by a timeout on the network link.

When the network link is having the status of LINK_UNSTABLE or SRV_UNAVAILABLE, the EPG recommendation needs to switch from server mode EPG recommender to mobile mode.

Referring to Fig. 1, when user queries EPG on a mobile device through its mobile GUI, the following operational flow may occur:

1. EPG query interface obtains the mobile working mode to decide which recommender to use. If the mobile is in HOME mode, the EPG query string and mobile user profile will be forwarded to message module on the EPG server and the mobile is waiting for the query result from EPG server.
2. When EPG server gets a query message, the server user profile gets updated with the synchronized mobile user profile, then the full scale EPG recommender returns the recommendation result to the message module.
3. The mobile message module gets the recommendation results from EPG server and sends it directly to user GUI module for displaying to the user.

4. If there is a timeout occurred when message module is waiting for the EPG query result, the network link enters into LINK_UNSTABLE state (network link status between mobile and server), which indicates EPG server becomes unavailable temporarily. Consequently, the mobile built-in EPG recommender is used together with mobile user profile to accomplish the recommendation task on board. In such a manner, the handoff from HOME to REMOTE (mobile) is realized.

5. If mobile is in REMOTE mode, the query string will be passed to built-in EPG recommender directly and the mobile user profile module is used to store user TV watching activities. When EPG server is available (i.e. mobile is in HOME mode), the mobile user profile will be synchronized to the EPG server as described previously. The handoff from REMOTE to HOME only occurs at the next EPG query, when the mobile working mode (HOME, REMOTE or OFFLINE) is checked. When implemented, the handoff operation should be unnoticeable to the user.

4 Prototype and Simulation

Among the mobile operating systems, the latest Android from Google seems to be the most eye-catching one. Whereas people are developing/testing their applications for a combination of 70-500 platforms/handsets that run rival operating system platforms, such as Windows Mobile, Symbian or the iPhone SDK, Google's Android reduces that to one. The developers can simply write an application that works on the Android emulator, which can work on any phone. Besides, Android platform has some potential advantages over other mobile operating systems:

1. Selection: Android platform promises to work on most newer smart phones, and more networks channels will sell Android phones. Many handset manufacturers, such as Samsung, Motorola, LG, and HTC, will all employ Android technology in the near future.

2. Open source: Google is following its open source tradition with Android, making the source code available to developers. This means that any programmer can create customized versions of the platform and applications that match the feel and look of Android.

3. Unlocked phones: Many carriers "lock" their phones and it is against their terms of service to unlock it. Android will allow users to install an unlocked operating system, which could lead to the switch of providers by the ending of binding contracts.

4. Google products: Android will support all Google services, such as Google Maps, Gmail, etc. Smart phones running Android will enable users much easier and smoother access to Google's other products, which may give Android phones an edge over other mobile operating systems.

Our prototype adopts Android SDK, and is developed on eclipse 3.3.2. Fig. 4 shows the major modules of EPG application as well as the components of the Android architecture [22]. In our experiments, we've used a laptop computer to simulate mobile device, which runs Android platform, and a PC with Windows XP operating system to simulate the server. Android SDK requires all applications implemented on Android platform to use the Java programming language. In our prototype, Java is only needed for Android mobile device, whereas the server can be implemented using any programming language. UDP is adopted as the network protocol for communications between mobile and server in order to simplify network complexity.

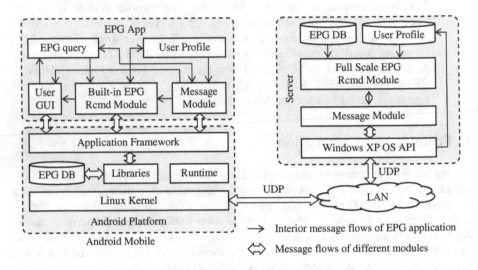

Fig. 4. EPG recommender application prototype

The recommender has been implemented using a Maximum Entropy classifier and this work has been described in detail in our previous paper [9]. On the mobile device, a simple Android UI is implemented as illustrated in Fig. 5. EPG query interface module and message module are also implemented as the core modules to realize the handoff operation. The EPG query interface module passes the query string to either built-in EPG recommender or EPG server based on mobile status. The message module deals with sending/receiving operation with timeout function provided by java SDK in case network is not reliable. To simulate the real case scenarios, a simple configuration UI on the mobile device is implemented as shown in Fig. 6, through which the timeout value and mobile status can be specified.

On the EPG server residing on a home network, a message module is implemented as a network server thread. When receiving an EPG query message from the remote mobile, the message module will forward the query to full scale EPG recommendation module. The two message modules on mobile and EPG server play an important role in the handoff operation. In such fashion, how the recommender core, whether for mobile or server, is implement is entirely independent of the framework prototype.

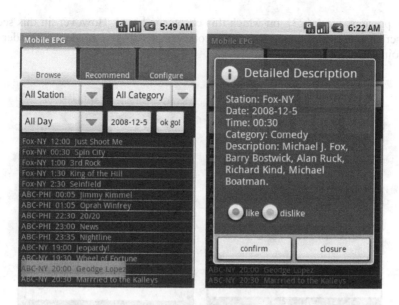

Fig. 5. Mobile EPG application GUI on Android platform

The handoff of EPG service is guaranteed seamless to the user via two mechanisms: the selection of time-out and buffer size for UDP message. The time-out is used to monitor the network return message and will trigger the network state to SRV_UNAVAILABLE thus the handoff of EPG service from server to mobile device. Currently, it is selected as 0.5 second without noticeable delay from the user. The buffer size for UDP message defines the maximum string length for a complete message. The buffer size is tuned to the EPG database used in the experiment and is capable of handling most of the recommendation tasks without switching of EPG service.

Finally, in our prototype, both ESG parser and EPG proxy for DVB-H and IPG parser and proxy for IPTV are implemented to support those two environments.

The feasibility of our proposed recommender handoff framework is tested in a LAN environment. To simulate the real wireless network, and test the auto handoff of EPG recommendation framework, the network port of EPG server is opened and closed at random time, while the mobile keeps sending EPG query to EPG server every several seconds. Through our experiment, we found that 0.5 second is an optimal time-out threshold, for a time out longer than 0.5s will make user noticing the delay or hand-off; whereas shorter than 0.5s will constantly trigger the built-in recommendation engine.

5 Related Work

The work described in this chapter crosses mainly with two domains: the TV recommendation and personalization; and mobile architecture related to mobile TV (or DVB-H). We have not found any literatures that marry these two domains except for our previous work [23], on which this chapter is based. However, in this section, we describe some representative related work which will give readers a broader view of ongoing efforts.

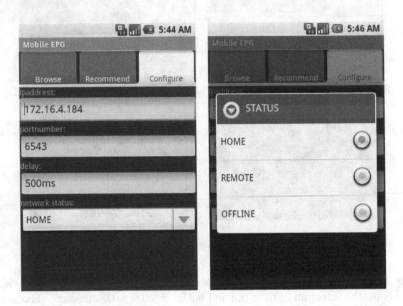

Fig. 6. Mobile EPG configuration GUI on Android platform

Various recommender approaches have been proposed based on EPG content, particularly category information. Blanco, et al [11] used TV ontology to build a collaborative recommendation system by defining similarity measures to quantify the similarity between a user's profile and the target programs. However, how to map the target program to the predefined categories is still a crucial problem, and in so-called TV ontology there is no acceptable current standard for the categories of TV programs.

Yu, et al [24] proposed an agent based system for program personalization under TV Anytime environment [25] using similarity measurement based on VSM (Vector Space Model). This work, however, assumes that the program information is available on a large storage media and does not address the problem of data sparseness and limited categories supported by most EPG providers. Pigeau, et al [26] presented a TV recommender system using fuzzy linguistic summarization technique to cope with both implicit and explicit user profile. This system largely depends on the quality of meta-data and solely on DVB-SI standard [27].

Jin Hatano, et al [28] proposed a content searching technique based on user metadata for EPG recommendation and filtering, in which four types of metadata are considered. The user attributes such as age, gender and place of residence are considered to implement an adaptive metadata retrieval technique. However, these attributes are too general for EPG recommendation. Personalized profiles for EPG recommendation mainly depend on users' interest or characteristics. Even though age and gender play certain roles, they are not the deciding factors.

L. Ardissono, et al [29][8] implemented a personalized recommendation of programs by relying on the integration of heterogeneous user modeling techniques. However, users cannot often declare their preferences in a precise way.

Among many TV recommendation and personalization work, explicit and implicit feedback as primary attributes to user profile have been largely used and studied. For example, Ehrmantraut, et al [30] and Gena [31] adopted both implicit and explicit feedback for personalized program guide. Cotter, et. al [14] describes an Internet based personalized TV program guide using an explicit profile and a collaborative approach.

In more recent work, Weiss, et al [13] described a TV recommendation system based on user profile and content filtering primarily based on actor and genre. Yet, their system was still under testing and no precision/recall of the recommendation system has been reported. Gadanho, et al [21] studied three different implicit user feedback strategies in the context of TV recommendation based on Naïve Bayes classifier, and reported no significant difference on the recommendation performance among different implicit feedback strategies as to how to select the most reliable feedback information to feed the learning algorithms. They have also reported to have achieved 75-85% precision and 12-39% recall, with the accuracy ranging between 67-74%. In contrast, we base our recommendation core engine on content based classification [19] with the precision and recall precision of 72-78%, recall of 58-75% and overall F1 ranging between 63% and 80%.

Some recommender systems attempted to integrate multiple prediction methods, such as neural network, given the user's reactions to the system's recommendations. One approach to merging multiple methods is to use relevance feedback technique, which can benefit from an informed tuning parameter. For example, Hiroshi Shinjo, et al [32] implemented an intelligent user interface based on multimodal dialog control for audio-visual systems, in which a TV program recommender based on viewing-history analysis is included. The profiler is built using groups of keywords analyzed from EPG content.

Errico, et al [33] have proposed a presence based collaborative recommender system for emerging networked AV display products with potential application to IPTV that access a variety of content sources over Internet and conventional content delivery channels. It proposed a recommendation system that has a social aspect enabled via the culmination of the current endorsements plus the previous endorsements of peers on the network. Their focus is slightly different from our work in this chapter but can be a complement to our system, in which social aspect and collaborative recommendation can be incorporated. What is missing from their paper though is the mobile aspect of the recommender.

More recently, Chen, etc al [34] have proposed a Community-based Program Recommendation (CPR) for the next generation EPG. It analyzed and categorized users with similar viewing habits into one community, and then recommended programs to users to help them quickly find the programs they want to watch. The deployment of CPR involves content providers and network operators, which makes the network topology more complex and the hardware more costly. Additionally, categorizing users with similar viewing habits into one community will weaken personalization level. Similarly, Tan, etc al [35] have constructed an EPG recommender framework with information exchange between users. The information exchange network is based on P2P with super nodes.

In mobile related work, Bostro, et al [36] has developed Capricorn that contains a number of embeddable widgets for installing on a mobile phone device including collaborative filtering widget. The collaborative filtering recommends frequently seen combination widgets and accordingly dim on/off appropriate widgets on the mobile phone. In one abstract level, this work is related to recommendation of services on a mobile phone, however, this system is not scalable as the need for more complex recommendations on a mobile phone increases because the computational resources on a mobile phone alone are too limited.

Jean-Paul M.G. Linnartz, et al [3] described the conception and development of a solution for improved mobile reception of the digital video broadcast standard for handheld (DVB-H). Their work included a simple characterization of mobility-induced interference based on a suitable model for the doubly selective channel. Their work, although related to DVB-H, did not involve mobile architecture or any upper level recomendation architecture.

Sungjoon Park, et al [15] implemented an agent-based personalized channel recommendation system at client-side in a mobile environment. The core module of the personalized channel recommendation system is Client Manager Agent, which limits the mobile application scenario to mobile stand-alone mode as compared to our system.

Wang Hui, et al [17] presented an architecture developed within the MING-T (Multistandard Integrated Network Convergence for Global Mobile and Broadcast Technologies) Project with the main purpose of experimenting an end-to-end solution to support innovative personal interactive multimedia services for convergent networks of DVB-H and 2G/3G. Their focus is on the general mobile application, which is a good candidate platform for our system.

Alan Brown, et al [18] discussed a novel architecture, based on the deployment of the SIP Service in a mobile OSGi framework, which effectively delivered PAN (Personal Area Network) and home network device interoperability. Combination of this architecture and our architecture described in this chapter can provide more real and practical system for the users.

6 Conclusion

Our proposed EPG recommendation framework encompasses an EPG recommender system that can be adapted to various network configurations. The recommendation framework can also be operated for both stand-alone EPG service on a mobile mode and full scale EPG service on a home server, or can be easily configured to handle the handoff of different recommendation services when a mobile user is roaming between inside and outside home.

The reference implementation has proved the network feasibility of the proposed framework. The handling of electronic program guide data from disparate sources such as DVB-H, IPTV, or the Internet is implemented via the definition of our core EPG database structure for the purpose of recommendation and EPG proxy for mapping any external data source to the core EPG data structure. Further, the handoff of recommendation tasks is realized by a simple definition of working mode of the mobile device and status of the network link. Particularly, the handling of network time-out and overflow of UDP buffer size guarantees the seamless handoff even during the middle of the execution of a recommendation task.

Finally, the recommendation framework is implemented on the new Android platform and is close to be deployed on a real mobile device.

Acknowledgement

Authors of this chapter wish to thank Anhui Wang, Wenliang Chen, Zhenxing Wang, and Huizhen Wang at the Institute of Computer Theory and Technology of Northeastern University for their support of this work.

References

1. Vare, J.: A prioritization method for handover algorithm in IPDC over DVB-H System. In: Proceeding of IEEE Int. Symposium on Broadband Multimedia Systems and Broadcasting, pp. 1–5. IEEE Press, Las Vegas (2008)
2. DVB-H standard, http://www.dvb-h.org
3. Linnartz, J.M.G., Filippi, A., Husen, S.A., Baggen, S.: Mobile reception of DVB-H: How to make it work? In: Proc. IEEE SPS-DARTS 2007, Belgium (2007)
4. Park, J., Blythe, M., Monk, A., Grayson, D.: Sharable digital TV: Relating ethnography to design through un-useless product suggestions. In: Proc. of CHI 2006, Canada, pp. 1199–1204 (2006)
5. Ma, M., Wilkes-Gibbs, D., Kaplan, A.: IDTV Broadcast Applications For a Handheld Device. In: IEEE International Conference of Communication (ICC), Paris, vol. 1, pp. 85–89 (2004)
6. Isobe, T., Fujiwara, M., Kaneta, H., Noriyoshi, U., Morita, T.: Development and features of a TV navigation system. J. IEEE Trans on Consumer Electronics 49(4), 1035–1042 (2003)
7. Takagi, T., Kasuya, S., Mukaidono, M., Yamaguchi, T.: Conceptual matching and its applications to selection of TV programs and BGMs. In: IEEE Int. Conf. On Systems, Man and Cybernetics, vol. 3, pp. 269–273. IEEE Press, Tokyo (1999)
8. Ardissono, L., Gena, C., Torasso, P.: User Modeling and Recommendation Techniques for Personalized Electronic Program Guides. In: Ardissono, L., Maybury, M.T., Kobsa, A. (eds.) Personalized Digital Television: Targeting Programs to Individual Viewers. Kluwer Academic Publishers, Dordrecht (2004)
9. Zhu, J., Ma, M., Guo, J.K., Wang, Z.: Content Classification for Electronic Programming Guide Recommendation for a Portable Device. Int. J. of Pattern Recognition and Artificial Intelligence 21(2), 375–395 (2007)
10. Chang, G., Zhu, C., Ma, M., Zhu, W., Zhu, J.: Implementing a SIP-based Device Communication Middleware for OSGi Framework with Extension to Wireless Networks. In: 1st International Multi-Symposiums on Computer and Computational Sciences, Hangzhou, pp. 603–610 (2006)
11. Blanco, Y., Pazos, J., Lopez, M., Gil, A., Ramos, M.: AVATAR: An Improved Solution for Personalized TV Based on Semantic Inference. J. IEEE Trans. on Consumer Electronics 52(1), 223–232 (2006)

12. Zhang, H., Zheng, S.: Personalized TV program recommendation based on TV-anytime metadata. In: Proceedings of the Ninth International Symposium on Consumer Electronics (ISCE 2005), Macao, pp. 242–246 (2005)
13. Weiss, D., Scheuerer, J., Wenleder, M., Erk, A., Gulbahar, M., Linnhoff-Popien, C.: A User Profile-based Personalization System for Digital Multimedia Content. In: Proc. of the 3rd International Conference on Digital Interactive Media in Entertainment and Arts (DIMEA), Greece, pp. 281–288 (2008)
14. Cotter, P., Smyth, B.: PTV: Intelligent Personalised TV Guides. In: Proc. of the 12th Innovative Applications of Artificial Intelligence (IAAI) Conference, Texas, pp. 957–964 (2000)
15. Park, S., Kang, S., Kim, Y.K.: A channel recommendation system in mobile environment. J. IEEE Transactions on Consumer Electronics 52(1), 33–39 (2006)
16. Xu, J., Zhang, L., Lu, H., Li, Y.: The development and prospect of personalized TV program recommendation systems. In: Proc. of the IEEE 4th Int. Symposium on Multimedia Software Engineering (MSE), California, pp. 82–89 (2002)
17. Wang, H., Song, Y.L., Tang, X.S., Zhang, P.: A General Architecture in Support of Personalized, Interactive Multimedia Services in the Mobile Broadcast Convergent Environment. In: 3rd International Conference on Testbeds and Research Infrastructure for the Development of Networks and Communities (TridentCom 2007), Florida, pp. 1–6 (2007)
18. Brown, A., Kolberg, M., Bushmitch, D., Lomako, G., Ma, M.: A SIP-based OSGi device communication service for mobile personal area networks. In: 3rd IEEE Consumer Communications and Networking Conference (CCNC 2006), Las Vegas, vol. 1, pp. 502–508 (2006)
19. Ma, M., Zhu, J., Guo, J.K.: A Recommender Framework for Electronic Programming Guide on A Mobile Device. In: Proceeding of IEEE Int. Conf. on Multimedia and Expo., Beijing, pp. 332–335 (2007)
20. Mann, G., Bernsteins, I.: Digital Television, Personal Video Recorders and Convergence in the Australian Home. In: Proc. of DIMEA 2007, Perth, Western Australia, pp. 121–130 (2007)
21. Gadanho, S., Lhuillier, N.: Addressing Uncertainty in Implicit Preferences: A Case-Study in TV Program Recommendations. In: Proc. of RecSys 2007, USA, pp. 97–104 (2007)
22. Open Handset Alliance, Android - An Open Handset Alliance Project, http://code.google.com/
23. Ma, M., Zhu, C., Tan, C., Chang, G.R., Zhu, J., An, Q.: A recommender handoff framework with DVB-H support on a mobile device. In: Proceeding of the First Int. Workshop on Mobile Multimedia Processing (WMMP 2008), Tampa, Florida, pp. 64–71 (2008)
24. Yu, Z., Zhou, X., Shi, X., Gu, J., Morel, A.: Design, implementation, and evaluation of an agent-based adaptive program personalization system. In: Proceedings of the IEEE 5th Int. Symposium on Multimedia Software Engineering, Taiwan, pp. 140–147 (2003)
25. TV Anytime Forum, http://www.tv-anytime.org
26. Pigeau, A., Raschia, G., Gelgon, M., Mouaddib, N., Saint-Paul, R.: A fuzzy linguistic summarization technique for TV recommender systems. In: Proceeding of the IEEE Int. Conf. on Fuzzy Systems, Missouri, pp. 743–748 (2003)
27. Specification for Service Information (SI) in DVB Systems, DVB Document A038 Rev. 1 (2000)
28. Jin, H., Kyotaro, H., Masahito, K., Katsuhiko, K.: Content Recommendation and Filtering Technology. J. NTT Technical Review 2(8), 63–67 (2004)

29. Ardissono, L., Gena, C., Torasso, P., Bellifemine, F., Chiarotto, A., Difino, A., Negro, B.: Personalized Recommendation of TV Programs. In: Cappelli, A., Turini, F. (eds.) AI*IA 2003. LNCS, vol. 2829, pp. 474–486. Springer, Heidelberg (2003)
30. Ehrmantraut, M., Herder, T., Wittig, H., Steinmetz, R.: The personal Electronic Program Guide- towards the pre-selection of individual TV programs. In: Proc. of CIKM 1996, Maryland, pp. 243–250 (1996)
31. Gena, C.: Designing TV viewer stereotypes for an Electronic Program Guide. In: Proceedings of the 8th International Conference on User Modeling, Germany, vol. 3, pp. 274–276 (2001)
32. Shinjo, H., Yamaguchi, U., Amano, A., Uchibe, K., Ishibashi, A., Kuwamoto, H.: Intelligent User Interface Based on Multimodel Dialog Control for Audio-visual Systems. Hitachi Review 55, 16–20 (2006)
33. Errico, J.H., Sezan, I.: Presence Based Collaborative Recommender for Networked Audiovisual Displays. In: Proc. of IUT 2006, Australia, pp. 297–299 (2006)
34. Chen, Y.C., Huang, H.C., Huang, Y.M.: Community-based Program Recommendation for the Next Generation Electronic Program Guide. J. IEEE Transactions on Consumer Electronics 55(2), 707–712 (2009)
35. Tan, C.G., Zhu, C., Wang, X.W., Chang, G.R.: An EPG Recommender Framework with Information Exchange between Users. In: Proceedings of the 5th International Conference on Hybrid Intelligent Systems, China, vol. 2, pp. 453–456 (2009)
36. Bostrom, F., Floreen, P., Liu, T., Nurmi, P., Oikarinen, T.K., Vetek, A., Boda, P.: Capricorn – An Intelligent Interface for Mobile Widgets. In: Proc. of IUI 2008, Spain, pp. 417–418 (2008)

MuZeeker: Adapting a Music Search Engine for Mobile Phones

Jakob Eg Larsen, Søren Halling, Magnús Sigurðsson, and Lars Kai Hansen

Technical University of Denmark,
Department of Informatics and Mathematical Modeling,
Richard Petersens Plads, Building 321
DK-2800 Kgs. Lyngby, Denmark
{jel,lkh}@imm.dtu.dk, {sch,mks}@muzeeker.com

Abstract. We describe MuZeeker, a search engine with domain knowledge based on Wikipedia. MuZeeker enables the user to refine a search in multiple steps by means of category selection. In the present version we focus on multimedia search related to music and we present two prototype search applications (web-based and mobile) and discuss the issues involved in adapting the search engine for mobile phones. A category based filtering approach enables the user to refine a search through relevance feedback by category selection instead of typing additional text, which is hypothesized to be an advantage in the mobile MuZeeker application. We report from two usability experiments using the think aloud protocol, in which N=20 participants performed tasks using MuZeeker and a customized Google search engine. In both experiments web-based and mobile user interfaces were used. The experiment shows that participants are capable of solving tasks slightly better using MuZeeker, while the "inexperienced" MuZeeker users perform slightly slower than experienced Google users. This was found in both the web-based and the mobile applications. It was found that task performance in the mobile search applications (MuZeeker and Google) was 2—2.5 times lower than the corresponding web-based search applications (MuZeeker and Google).

Keywords: search engine, music, user interfaces, usability, mobile.

1 Introduction

The rapid growth of multimedia data available to the Internet user underlines the importance and complexity of information retrieval. The amount and diversity of data and distributed database structures challenge the way search engines rank their results and often several million results are presented to a single query. Furthermore, the constraints on mobile devices in terms of limited display size and resolution along with limited text entry facilities makes information search and retrieval an even bigger challenge. The usability of search engines on mobile devices is essential in terms of optimizing and utilizing the screen real-estate and limiting the amount of explicit text input required by the end-user [1].

X. Jiang, M.Y. Ma, and C.W. Chen (Eds.): WMMP 2008, LNCS 5960, pp. 154–169, 2010.

We introduce the Zeeker Search Framework [2] which uses Wikipedia [3] to generate categorization of the individual search results, thus enabling search refining by a category selection instead of typing a new query. The information architecture of Wikipedia is a potentially useful knowledge base for machine learning and natural language processing and can be used to empower the Zeeker Search Framework's contextual understanding and ability to categorize search results. A better understanding of the semantics of user queries and web content by categorization facilitates presentation in mobile devices and usability. We hypothesize that Wikipedia can offer the necessary categorization in many domains including the domain of interest here, namely music.

2   Search Engine

Wikipedia is a well known on-line free encyclopedia characterized by the fact that anyone can edit articles, it has tremendous support, and the number of articles is increasing [3]. Wikipedia users co-author and edit articles on topics where they believe to have factual and neutral insight. The Wikipedia collective intelligence assures that articles are labeled and categorized by many editors thus reducing problems with idiosyncrasy that may haunt knowledge bases established by individuals or specific organizations.

The Zeeker Search Framework [2] [10] (ZSF) is aimed to be flexible offering "real-time" specialization through Wikipedia. In many domains the broad group of contributors warrants swift updates in reaction to news and domain events. The framework is developed primarily with search result categorization, topic specialization and result presentation in mind. The search engine in the ZSF is based on the well-known and widely used vector space model where the structure can be split into three different processes, namely Document Processing, Data Indexing and Query Processing.

2.1   Document Processing

The document processing part is responsible for normalizing the input documents and splitting them up into indexable units, i.e. terms. When breaking text up into terms, the most obvious way is to break on white spaces – i.e. define a single word as a term. This approach is called bag-of-words where compositional semantics are disregarded [4]. ZSF uses the bag-of-words approach. Other language features such as punctuations, digits, dates and text casing are also handled in ZSF.

Stop word removal is implemented in the ZSF and is based on the assumption that if a term occurs in more than 80% of the documents it should be considered a stop word [5]. Stop words are pre-calculated in a stop word list. Stop words cannot be removed uncritically and must be based on the vocabulary and context of the documents [6].

Stemming is used to remove word suffixes (and possibly prefixes). This reduces the number of unique terms and gives a user's search query a better recall. Several stemming algorithms exist whereas ZSF makes use of the well known Porter stemmer due to its simplicity and elegance while still yielding comparable results to more sophisticated algorithms.

2.2 Data Indexing

A naïve indexing approach is to simply index every term that occurs within the documents. This would require enormous amounts of data storage, and could result in computational bottlenecks and slow response times to queries.

The indexed terms are stored in an inverted file. An inverted file is a file structure representing the relation between words and their locations in a document or a set of documents. The inverted index [7] has many variants but usually contains a reference to documents for each word and, possibly, also the position of the word in the document. Basically, the inverted file contains a line for each word where the line starts with the word then followed by the list of documents containing the word. Furthermore, if word positions are also stored in the inverted file, then the each line would start with a word, followed by a list of tuples wherein the document number and word position would be stored. For example the line:

computer   (1,3), (2,3)

indicates that the word 'computer' occurs in document 1 and document 2 at position 3 in both documents. With term positions stored as well as the document number, term proximity can be weighted into the queries. ZSF uses a full-inverted index with term positions stored as well.

In order to reduce the number of terms indexed and the general size of the index, ZSF makes use of lexical analysis in the form of Part-Of-Speech (POS) taggers. POS taggers analyze sentences and tag terms with their syntactical groups such as verbs, nouns, numbers, punctuations etc. When using POS taggers it has been shown that the vocabulary (term histogram) can be reduced by up to 98% [8], depending on the vocabulary. With a moderate POS filtering strategy, stopping and stemming it was possible to reduce the index size for a Wikipedia test set by 14% in unique terms and a decrease of 17% in disk space usage.

2.3 Query Processing

Before a query can be submitted to the search engine, the query terms should be represented in the same way as the documents are represented in the index, i.e. using the vector space model. Terms in the query vector are normalized like index terms, i.e., by stop word removal, stemming etc.

With documents and queries represented as high-dimensional vectors using the vector space model, the matching documents are found using the cosine similarity measure with documents normalized and term-weighting applied. The Cosine Similarity Measure is calculated in the following way for the query vector q and the document vector d:

$$\cos \theta = (q \lozenge d)/(|q||d|)$$

This means the smaller the angle between the vectors, the more similar they are where $\cos \theta = 1$ means that vectors are identical.

Since the vector space model with cosine similarity is used for retrieval, the ranking of the results becomes fairly simple. The ranking of documents is done simply by the value of the similarity measure between the query vector and document vectors. The higher the value, the higher the document will appear in the list of results. Categories are indexed and retrieved from a special category-index where the ranking is also done using the cosine similarity measure.

2.4 Zeeker Search Framework Architecture

The ZSF structure is divided into three separate layers as shown in Fig. 1, namely a data layer, an XML web service and a presentation layer. The layers enhance the search engine's flexibility, making it easier to improve and expand across platforms and technologies.

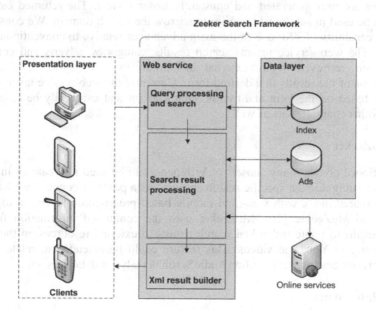

Fig. 1. Overview of the Zeeker three-tier system architecture: presentation, web service and data layer. A standard web service interface links the presentation and framework.

The data layer contains the document and category indexes, associated adverts and any additional meta-information such as query suggestions, document types etc. Customized indexes can be created using any subset of articles from the Wikipedia knowledge base. This is achieved by indexing articles related only to a specific category along with its sub-categories and articles, e.g., music, birds, movies, wine, history, etc.

Besides deep categorization and relatively high trust article contexts [9], Wikipedia has a number of additional useful attributes. All articles have a quality measure, such as "stub", "normal" and "featured" that indicates how Wikipedia rates the context of the article. Every article also has associated discussion pages and log files of what has changed, who changed it and when the change occurred. These article attributes provide information, e.g. about the article quality. Long discussion pages, many major and minor edits, the profile of the user responsible for the editing, amount of links to other Wikipedia articles, amount of external links, how many images are included and which images, templates used etc. could potentially be used as quality indicators for a specific article. These measures can also be used to implement filters to remove poor quality articles.

The XML web service layer provides the search and information retrieval service. Using the knowledge gained from Wikipedia, the ZSF is able to classify a given user submitted query to a set of ranked categories along with a set of ranked search results. In contrast to other datamining and clustering based search engines the Wikipedia categories are user generated and constantly under review. The returned categories can then be used in subsequent queries to narrow the search domain. We consider the filtering capability of ZSF one of the main advantages relative to conventional search engines. The web service returns search results, categories, adverts and contextual information retrieved from external resources in XML format, facilitating presentation of the results in a desired form. Currently the web service layer supports searches based on the latin and cyrillic character set and can easily be extended to support other character sets as well.

2.5 MuZeeker

As mentioned above, many subsets of Wikipedia can be used to create an index and thereby creating domain specific search engines. As a proof-of-concept, we have built a music search index with web- and mobile-based presentation layers on top of the ZSF called MuZeeker [10]. MuZeeker uses the contextual information from the search results to relate individual search results to external resources, in the current implementation YouTube videos. This feature could be extended to relate to other external resources such as Last.fm (AudioScrobbler), lyrics databases, etc.

2.6 Related Work

The encyclopedic knowledge available in Wikipedia has been deployed in other contexts besides ZSF such as [11] where Wikipedia data is used for semantic disambiguation and recognition of named entities. Semantic information based on Wikipedia has also been used in search engines such as Powerset [12] and Koru [13]. Powerset uses its semantic understanding to find answers to user's questions within the articles while Koru uses its semantic understanding to expand queries and guide the user interactively to the best results. Other search engines using Wikipedia data include Wikiwix [14] and Exalead [15]. These two engines associate categories with each search result but do not provide category searches at the same level as ZSF.

3 User Interfaces

Two prototype user interfaces (the presentation layer) have been developed for the MuZeeker search engine. The web-based user interface for MuZeeker is designed with an appearance similar to web search engines such as Google (Fig.2), whereas the mobile user interface is tailored for mobile phones.

3.1 Queries and Search Results

In both interfaces the user can key in the query terms in an input field, which can also contain a set of parameters to refine the query. The syntax of the query is similar to that of typical web search engines, that is, it includes a number of search terms and supports a basic set of operators: AND, OR, NOT, EXACT (match), and ORDER. These operators can be used along with other operators, thus giving the user a flexible syntax to build the queries needed to find the most relevant information. A basic search query consists of a number of search terms, which would match a text where all the search terms are available, as by default, the search engine uses a boolean AND query of all the query terms. Category filters can be added using a "-cat" parameter to the search query and several categories can be included.

The primary difference between MuZeeker and conventional search engines is that MuZeeker returns two sets of results for any search query. One is a set of *articles* matching the search query, and the other is a set of matching Wikipedia *categories*. The set of matching categories enables the user to refine his/her search – a kind of relevance feedback.

The user can either add categories by including the "-cat" parameter mentioned above, or by selecting one of the matching categories returned by the search engine. Selecting a category is equivalent to manually adding the category to the search query and submitting it. Users repeat this process until a satisfying result has been obtained. A result will include a snippet from the original Wikipedia article (the first couple of lines), a link to the original Wikipedia article and may include other resources as well.

3.2 Web-Based User Interface

The web-based user interface for MuZeeker is designed similarly to conventional search engines, while allowing for the additional features present. For current design see Fig. 2 and [16]. A ranked list of search results is linked to additional information in the web interface.

In addition to the Wikipedia search results, additional information about each search result is included when available. This information is primarily the document type of the search results, e.g. musical artist, television series, mountain, software etc. When the information is available, a link to a video search on YouTube is provided. In MuZeeker the links could be to a music video related to an article describing a particular song.

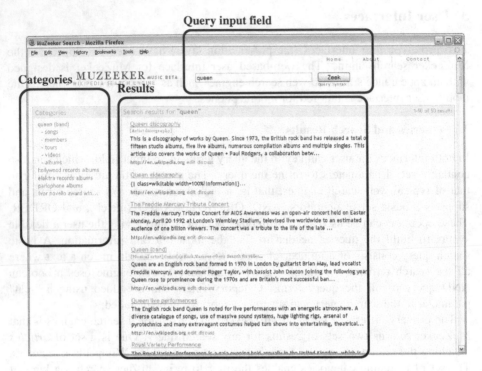

Fig. 2. The web-based user interface for the MuZeeker search engine. The area in the top provides the conventional search *query input field*, whereas the bottom part with search results is divided into two parts: matching *categories* and matching *results*.

3.3 Mobile User Interface

Where the web-based user interface for the MuZeeker search engine is fairly trivial the user interface for the mobile devices imposes a set of challenges due to limited screen real estate and limited input capabilities. Thus the focus in this study has been on adapting the search engine user interface for mobile platforms, such as mobile phones.

The user interface for the mobile MuZeeker search application is adapted for the S60 mobile platform, see Fig. 3. The mobile application is implemented for Nokia N95 and N95 8GB mobile phones using the Web-Runtime (WRT) environment [17].

Due to the limitations of the platform and devices the user interface has been optimized. In particular the challenge is to adapt the amount of information as present in the web-based user interface (Fig. 2), without sacrificing ease of navigation in the mobile user interface. The core functions included are the search facility, the search results, the category selection and the display of a small snippet of the articles. Links to the original Wikipedia article and YouTube video are included in the present prototype. Selecting a link will launch external applications on the mobile device in order to display the content (mobile web browser or *realPlayer*). The constraints of the display allow only about ten lines of text on the screen. This makes it difficult to include both the list of search results and the list of categories on the same screen and still allow easy user navigation. Therefore, the search results and categories have been divided into two screens, as illustrated in Fig. 3.

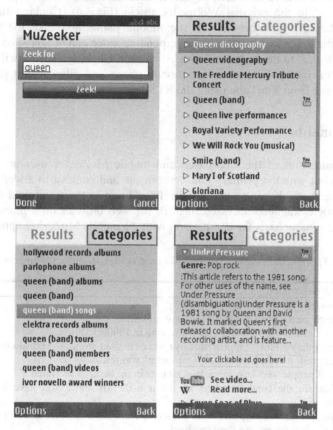

Fig. 3. Four screenshots from the MuZeeker mobile application prototype: a) main search interface, b) ranked list of search results, c) ranked list of categories and d) expanded result

By default the ranked list of search results is shown to the user, while shifting to the ranked list of categories is done by pressing one button (the *right* button on the 4-way navigation key). The number of search results and categories displayed is ten items due to the screen limitations to avoid scrolling a list, which would likely be perceived as cumbersome. The intention of the approach is that text entry should be limited to the extent possible. Thus, a typical scenario is that the user types in some search terms using the main search interface (Fig. 3.a). By using the results (Fig. 3.b) and categories (Fig. 3.c) the user can refine the search by adding a category to the search query by *selecting* the category from the list (Fig. 3.c). The advantage is that little typing (text input) is required in order to refine the search, as it can be done using the four-way navigation key. Nevertheless, the mobile application does support the full query syntax and operators as described above, but it would typically be too cumbersome to use, due to the text entry limitations on standard mobile phones. Each item in the list of search results (Fig. 3.b) can be expanded as shown in Fig. 3.d. This provides the first few lines of the article from Wikipedia along with genre information as indexed in the MuZeeker search engine. In addition each result item provides a link "Read more..." to the original Wikipedia article (Fig.3.d). Clicking the link will launch the built in web browser in the mobile phone and take the user to the Wikipedia article. For songs where a corresponding video can be found on YouTube a link "See video..." to this video is provided as well. Clicking the link will launch the built in *realPlayer* application in the mobile phone and allow video playback via direct streaming from YouTube using the RTSP protocol.

4   Experiments

Usability evaluations of the web-based and mobile MuZeeker user interfaces have been carried out with the primary goal to compare and contrast to a well-established search engine, here chosen to be Google. In addition to the traditional Google web interface Google was also accessed through the web browser on a mobile phone in order to have a better basis for comparison with the mobile MuZeeker application.

Table 1. Overview of the 15 tasks used in the usability experiments

Task	Task description
1	Who left "The Supremes" in 1970?
2	Which members of the band U2 have nicknames?
3	Who did a cover version of a Prince song in 1990?
4	How many albums (approximately) were released from the record company the same year that the album "OK Computer" was released?
5	List 3 people who have been lead singers in the band "Van Halen"?
6	What are the band members of the non-Danish music you have heard most recently?
7	What do "Britney Spears" and the composer "John Adams" have in common? HINT: It is something they received.
8	Take a small break and then continue with the remaining tasks using the other search engine.
9	Which album (that was later made into a movie) did "Pink Floyd" release in the late 70's – early 80's?
10	Who produced the Michael Jackson album "Thriller"?
11	List three artists what have made a cover version of the song "Blowing' In The Wind".
12	The composer "Franz Liszt" wrote music in the 19's century. Which of his fellow countrymen has written similar music?
13	Find information about the non-Danish music you have heard about (online/offline), such as the title of the latest album or names of the band members.
14	Find song titles, artist and album title for the music played in the radio right now.
15	What do "Foo Fighters", "Nirvana" and "Queens of the Stone Age" have in common?

The evaluations were implemented so that participants solved 15 different tasks involving retrieval of music related information using the two different search engines. The tasks used in the experiments are shown in Table 1. Observe that task 8 is a short break to switch search engine. This means that 7 different music related tasks were used with each search engine during an experiment. Moreover it should be observed that the tasks 6, 13 and 14 were open, as they relate to the individual test participant. These tasks were included to observe the participants performing tasks that would involve music they could relate to.

Testing and comparing MuZeeker with a general search engine is challenging since MuZeeker is domain specific. Therefore, a Custom Google Search Engine [18] was created to make the results comparable. This approach was used in both usability experiments. The Custom Google Search Engine was designed to only search within the *en.wikipedia.org* domain excluding sites having the character ':' in the url and included a default search term – music – which was invisible to the test participants. This means that in the experiments, Google's functionality was the same as the standard Google search engine, but a set of predefined parameters were automatically added in each search. Furthermore, to assure that it was the search engines that were tested (and not Wikipedia), test participants were instructed not to use links pointing out of the search engine domain, i.e., solving the tasks using only the text snippets.

The experiments were conducted using the think aloud protocol as described in [19] with N=20 participants in total. The first experiment (N=10 participants) compared the web-based MuZeeker application with the web-based Google interface. The second experiment (N=10 participants) compared the mobile MuZeeker application with Google used on the same mobile phone (Nokia N95). Both experiments had nine male and one female. In the experiment with the web-based interface the first two tests were considered pilot tests which led to protocol adjustments. The average age in the web-based experiment was 26.6 years and all except one had experience in designing or programming software and all had, or were currently studying at MSc or PhD level. The average age in the mobile experiment was 23.8 years and all had experience in designing or programming software and all were currently studying at MSc level. In both experiments all ten participants reported Google as their preferred web search engine.

The tasks were designed without precise prior knowledge of how difficult they would be to solve in the two search engines, however, with an intended increase in difficulty. Some tasks were closed tasks such as task 1: *Who left "The Supremes" in 1970* while others were open and thereby yielded a broader test. The open tests have been left out of the comparisons below, because different search objectives results in different completion times and ratios. The latter tests in each search engine were complicated tests which asked the test participants to find similarities between two persons, bands or similar.

Information was gathered about participants' age, education, and general use of the Internet, search engines, and music. A brief introduction was given. Especially in the experiment with the mobile application it was necessary to give instructions to the mobile phone, as the participants were not familiar with the particular mobile phone used in the experiment. Moreover a brief introduction to the mobile applications was provided. The 15 tasks were given sequentially in writing and each test participant was timed, and asked to read aloud each task. They were told that some tasks might not be solvable, which is believed to make test participants behave closer to a typical situation, in which, users are not aware whether or not a given task can be solved (answer can be found using the search engine). Answers were regarded as correct, incorrect or partially correct. As an example only the answer "Quincy Jones" was regarded as the correct answer to task 10. Answering both "Bono" and "the Edge" would be regarded as a correct answer to task 2, whereas mentioning just one of them would be regarded partially correct, and none of them incorrect. Task 15 has several possible answers and the answer is regarded correct if the participant mentions just one of the possible answers: "Share the band member Dave Grohl", "Based in Seattle", or "play same genre of music".

In the concluding debriefing session, the test participants were asked to rate their experience using the two different search engines in order to determine their subjective satisfaction. The ratings were based on the measurement tool Questionnaire for User Interface Satisfaction (QUIS) [20] using the following parameters to assess the participants' perception of and subjective satisfaction using the interfaces of the two search engines: Terrible–Wonderful, Frustrating–Satisfying, Dull–Stimulating, Confusing–Clear and Rigid–Flexible. The test participants also indicated which search engine they would like to use for subsequent similar tasks and to rate the satisfaction with the tasks they were given in terms of the intervals: Hard – Easy, Hard to grasp – Easy to grasp and Unrealistic – Realistic.

5   Results

On average, MuZeeker performed better than the custom Google search engine in terms of the percentage of correctly solved tasks, although only marginally as can be seen in Fig. 4.

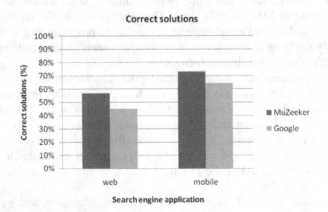

Fig. 4. Percentage of correct solutions measured in the two usability experiments (web and mobile) using the MuZeeker and Google search engine applications respectively

This was found in both usability experiments, that is, both when comparing the web-based applications and when comparing the mobile applications. It was also found that the percentage of correct solutions was slightly higher in the mobile experiments compared to the web-based experiments. However, it should be observed that on average considerably longer time was spent solving the tasks when using the mobile applications compared to the web-based applications. Therefore the amount of time spent solving the tasks could have influenced the percentage of correct answers.

On average, the custom Google search engines performed better than the MuZeeker applications in terms of the amount of time spent on solving the tasks, as can be seen in Fig. 5. This was found in both the web-based and mobile usability experiment, although the difference was only marginal in the web-based applications. These results indicate that "inexperienced" MuZeeker users are slower but on the other hand able to solve tasks slightly better using the MuZeeker applications compared to the custom Google search engines used.

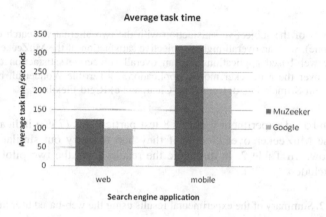

Fig. 5. Average time used for correctly solving tasks as measured in the two usability experiments (web and mobile) using the MuZeeker and Google search engine applications respectively

Participants generally found the tasks easy to grasp/understand and they also found the tasks realistic meaning that they could see themselves doing similar tasks in search engines. However, a majority of the test participants found that some of the tasks were hard to solve. In the experiment with the web-based applications the overall subjective satisfaction of MuZeeker was slightly higher compared to Google. In the experiment with the mobile applications the overall subjective satisfaction of the Google application was higher compared to the MuZeeker application, as can be seen in Fig. 6.

It is worth observing that the assessment of MuZeeker might be affected by problems experienced by participants using the mobile phone and application. This included challenges learning the features of the mobile application. Moreover, as the mobile MuZeeker application is a research prototype there are a few issues leading to potential usability problems. Some participants had initial problems focusing the input field and hitting the search ("zeek") button (Fig 3.a). A time consuming issue in the present prototype is the fact that it can be difficult to cancel an ongoing search. In a few cases it may lead to an application lock, which requires the user to restart the application. Finally if the user accidentally hits the "end call" button on the phone the application will exit and require a restart of the application. Similar problems were not present when using Google search through the mobile web browser on the mobile phone.

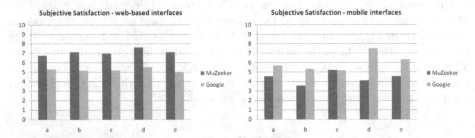

Fig. 6. Overview of the subjective satisfaction with the two alternative search engines (web-based and mobile), with an overall higher subjective satisfaction of the MuZeeker compared to Google in the web-based applications and an overall subjective satisfaction of the Google mobile search over the MuZeeker mobile application. a) Terrible-Wonderful, b) Frustrating-Satisfying, c) Dull-Stimulating, d) Confusing-Clear, and e) Rigid-Flexible.

In the web-based experiment 6 out of 8 test participants (75%) indicated that they would choose MuZeeker over Google if they had to carry out similar tasks in the future, as shown in Table 2. In this table the results from the two pilot experiments have been excluded.

Table 2. Summary of the experimental results using the web-based user interface

Web-based user interface	MuZeeker	Google
Tasks solved correctly (average)	56.8%	45.5%
Average completion time (correctly solved) (s)	124.0	99.9
Subjective preference	6	2

To the same question 5 out of 10 test participants (50%) indicated that they would choose MuZeeker over Google in the mobile experiment. A summary of the results are shown in Table 3.

Table 3. Summary of the experimental results using the mobile user interface

Mobile user interface	MuZeeker	Google
Tasks solved correctly (average)	72,2 %	64,2%
Average completion time (correctly solved) (s)	320	206
Subjective preference	5	5

It was observed in the experiments that the test participants were able to use the category feature in MuZeeker to solve the tasks. However, on inquiry it was found that none of the test participants were able to accurately explain what the category filter did and how they did it, although many said that they liked its results. Many test participants were surprised at how difficult it was to obtain certainty of the correctness of the snippet information, as they were presented out of context. Still some participants, although aware of these uncertainties, mistakenly made incorrect inferences solely out of the snippet information. Some even did this, although it contradicted what they thought they knew about a band or artist.

6 Discussion

Our observation was that MuZeeker as an application appears to be better suited for mobile phones targeted for music playing and video playback. The domain of the application – music search and playback – is in line with the applications offered by modern mobile phones and is therefore closer to current mobile phone usage patterns. It was observed during the experiments that the participants in both the web-based and the mobile usability experiments actively used the category selection feature of MuZeeker in order to solve the tasks.

The web-based user interface allows for more information to be displayed, and thus provides a better overview of search results. For example, along with each result a snippet of the resulting article is shown in the web interface (see Fig. 2), thus providing the user with cues whether the needed information is likely to be found in that search result. Due to the display limitations on the mobile device only the document title (or sometimes only a part of the document title) is shown. Either the user of the mobile user interface has to guess based on the title (or part of the title) or has to go through the tedious process of expanding each search result to get similar cues (as illustrated in Fig. 3.d). This involves a series of keypad clicks on the mobile device, whereas the user can typically accomplish the same task in the web-based user interface by simply skimming a single web page with 7—10 results visible at a time. The web-based user interface provides 50 results on a single (scrollable) web page, whereas the mobile is limited to 10, due to the display limitations.

By default expanding all search results in the mobile user interface would on the other hand only allow 1—2 result items to be visible at a time, which would force the user to scroll down to view the entire list. In the current solution 10 search results are visible and initially collapsed. Typically this leads to a two phase process when browsing the results, meaning that the user first judges the relevance of search results based on the titles, and second must manually expand relevant results to get further cues from the snippet of information shown in the expanded result. The usability of these solutions of course depends on the ranking ability of the search engine. If the information is available in the first couple of results shown this is less of a problem.

It was evident from the experiments that the mobile user interface has an impact on the average task performance. The results of the usability experiments indicate that participants took 2—2.5 times longer to perform the tasks using the mobile applications. This was the case for both the MuZeeker application and using Google search on the mobile phone. The slower performance obviously come from the user interface aspects explained above along slower typing on the mobile device (on the numeric keypad), less network bandwidth meaning that it takes longer to get the results of a search from the server, and also that participants were not familiar with the mobile phone used in the experiment. In addition part of the longer task time is due to some participants having initial problems using the mobile application, due to the prototype application issues mentioned previously. This included the initial learning of input field focus and hitting the search button, along with issues of the application locking or even accidental exit of the application in some cases during the experiments. The time spend on those issues were not subtracted from the overall task time. Due to the somewhat longer task performance time the entire mobile experiment of course took somewhat longer to carry out, compared to the experiments with the web-based applications.

Finally it is observed, that the experiment has taken a conservative approach to the comparison of the two search engines, MuZeeker and Google, as all test participants were experienced Google search users, but inexperienced MuZeeker users. The initial learning curve in MuZeeker was not subtracted in the time measurements of task completion. As such the experiment considers whether a novel search approach (including its initial learning curve) can compete with Google search for domain specific search, such as in the present case music.

7 Conclusions

We have described our Wikipedia domain specialized search engine MuZeeker. The user can refine a search in multiple steps by means of selecting categories. In this first version the domain supported by the MuZeeker search engine has been 'music'. We have carried out two initial experiments using two search application prototypes, web-based and mobile respectively. The experiments were carried out using the think aloud protocol with a total of 20 participants performing tasks using respectively MuZeeker and a custom Google search engine. The first experiment was carried out using web-based search applications and the second experiment using mobile search applications. In both experiments MuZeeker and Google were compared.

Our findings from experiments present evidence that MuZeeker users were capable of solving tasks slightly better using MuZeeker, but the "inexperienced" MuZeeker users performed slower than experienced Google users. Similar results were observed in the web-based and mobile experiments However, further studies are necessary in order to arrive at solid conclusions. 75% of the participants reported a subjective preference for the web-based MuZeeker application compared to Google for solving similar music information retrieval tasks. However, preferences were equal for the MuZeeker mobile application and the Google mobile search.

The category-based filtering approach enables users to refine a search by performing selections rather than typing additional text. This is highly useful in the mobile application where the amount of necessary text input is minimized and navigation support in refining a search is enhanced. The experiments also showed that the participants had problems explaining the category filters. In some cases the multiple steps made it unclear to the participants which query the shown search results related to. Our initial experiments with the mobile MuZeeker application demonstrated the search approach as a promising way to make large quantities of information searchable on mobile devices. However, further studies are needed in order to understand the search strategies chosen by the participants and the use of the category based filtering. Studies of other domains of interest are also relevant.

Acknowledgments. This project was supported by The Danish Research Council for Technology and Production, through the framework project 'Intelligent Sound', www.intelligentsound.org (STVF No. 26-04-0092). We would like to thank Forum Nokia for the support of mobile devices. Finally we would also like to thank the 20 participants that took part in the usability experiments.

References

1. Kamvar, M., Baluja, S.: Deciphering Trends in Mobile Search. IEEE Computer 40(8), 58–62 (2007)
2. Sigurdsson, M., Halling, S.C.: A topic-based search engine. Technical University of Denmark, IMM-Thesis-2007-91 (2007), http://orbit.dtu.dk/
3. Wikipedia, http://en.wikipedia.org
4. Sebastiani, F.: Machine learning in automated text categorization. ACM Comput. Surv. 34(1), 1–47 (2002)
5. Baeza-Yates, R., Ribeiro-Neto, B.: Modern Information Retrieval. Addison-Wesley, Reading (1999)
6. Frakes, W.B., Baeza-Yates, R.: Information Retrieval: Data Structures and Algorithms. Prentice Hall, Englewood Cliffs (1992)
7. Zobel, J., Moffat, A.: Inverted files for text search engines. ACM Comp. Surv. 38(2), 6 (2006)
8. Madsen, R.E., Sigurdsson, S., Hansen, L.K., Larsen, J.: Pruning the vocabulary for better context recognition. In: Proceedings of the IEEE International Joint Conference on Neural Networks, vol. 2, pp. 1439–1444 (2004)
9. Giles, J.: Internet encyclopedias go head to head. Nature 438, 900–901 (2005)
10. Halling, S., Sigurðsson, M., Larsen, J.E., Knudsen, S., Hansen, L.K.: MuZeeker: a domain specific Wikipedia-based search engine. In: Proceedings of the First International Workshop on Mobile Multimedia Processing (WMMP2008), Florida (December 2008)
11. Cucerzan, S.: Large-Scale Named Entity Disambiguation Based on Wikipedia Data. In: Proceedings of EMNLP-CoNLL 2007, pp. 708–716 (2007)
12. PowerSet, http://www.powerset.com
13. Milne, D.N., Witten, I.H., Nichols, D.M.: A knowledge-based search engine powered by wikipedia. In: CIKM 2007: Proceedings of the sixteenth ACM Conference on information and knowledge management, pp. 445–454 (2007)
14. Wikiwix, http://www.wikiwix.com
15. Exalead, http://www.exalead.com
16. MuZeeker search engine website, http://www.muzeeker.com
17. Widgets: Nokia Web Runtime (WRT) platform, http://www.forum.nokia.com/main/resources/technologies/browsing/widgets.html
18. Google Custom Search Engine, http://google.com/cse
19. Boren, M.T., Ramey, J.: Thinking Aloud: Reconciling Theory and Practice. IEEE Transactions on Professional Communication 43(3), 261–278 (2000)
20. Chin, J.P., Diehl, V.A., Norman, K.L.: Development of an instrument measuring user satisfaction of the human-computer interface. In: CHI 1988: Proceedings of the SIGCHI conference on Human factors in computing systems, pp. 213–218 (1988)

A Server-Assisted Approach for Mobile-Phone Games

Ivica Arsov, Marius Preda, and Françoise Preteux

Institut TELECOM / TELECOM SudParis/ARTEMIS,
91011 Evry, France
{Ivica.Arsov,Marius.Preda,Francoise.Preteux}@it-sudparis.eu

Abstract. While the processing power of mobile devices is continuously increasing, the network bandwidth is getting larger and the latency is going down, the heterogeneity of terminals with respect to operating systems, hardware and software makes it impossible to massively deploy mobile games. We propose a novel architecture consisting of (1) maintaining on the terminal only standardized operations (2) using a remote server for the game's logic and (3) updating the local scene by using compact commands. By identifying the game tasks that should remain local, we first analyze the performances of MPEG-4 3D graphics standard. We implemented an MPEG-4 player able to decode and visualize 3D and extended it for handling game content. We propose a client-server architecture and a communication protocol that ensure similar user experience while using a standard player - expected to be available on mobile devices in the near future.

Keywords: 3D graphics, mobile games, multimedia standard, client-server architecture.

1 Introduction

Mobile devices could not fail to be included in the ongoing growth of 3D graphics in the digital multimedia world. However, this trend does not satisfy expectations. Some studies published in 2001 forecast for 2007 a mobile games market of $18 billion, while the real size was only $2.5 billion. Different reasons were proposed to explain the "underperformance". Beside those referring to business maturity and usability, the main technological barrier was identified as the fragmentation of the mobile phones market, due to the heterogeneity of devices. In order to build a mobile application, a developer has to consider differences in hardware configurations and performances, operating systems, screen sizes and input controls. This diversity adds time and cost to the development cycle of a 3D application. To minimize these, several attempts at stabilizing the technology at different levels are currently in progress. At the hardware level, Nokia proposes, with the N-Series and particularly with the N-Gage, a relatively powerful terminal, with the intention of becoming a dedicated mobile console for games and graphics applications. Recently, Apple has chosen a similar strategy, with greater impact, by proposing the iPhone. At a higher level, but still in

X. Jiang, M.Y. Ma, and C.W. Chen (Eds.): WMMP 2008, LNCS 5960, pp. 170–187, 2010.

the hardware realm, stabilization is ongoing for graphics chipsets. Major actors from the traditional desktop graphics chipsets market, such as ATI and NVIDIA, propose low-consumption chips ready to be integrated into mobile devices.

On the other hand, several layers of software are available, such as low-level graphics libraries (OpenGL ES, Direct3D mobile), virtual machines (Java with JSR-184, a.k.a. M3G), and dedicated rendering engines (Mascot Capsule by HI Corporation or X-Forge by Acrodea). Despite all the efforts at stabilizing the technology mentioned above, we are still far from the situation in the PC world where there are fewer variables. In this case, a game has some minimum requirements with respect to hardware but it is generally independent of the manufacturer; the software layer is dominated by one OS, and two low-levels graphics libraries (DirectX and OpenGL) are widely adopted.

It is interesting to note that the current evolution of mobile phones follows a similar approach to the evolution of personal computers twenty years ago. However, the network performances available now may change the outcome. In the early nineties, two main trends were present on the PC market: Microsoft focused on increasing power of end-user terminals and their standardization while Oracle concentrated on server-centric software. Oracle invented the term "thin client", in which a dedicated server, able to run powerfull application and stream visualization commands to the client, assists the terminal. The advantages of this include lower administration costs, increased security and more efficient use of computing resources.

In addition, efficient protocols such as Independent Computing Architecture (ICA) by Citrix or NX by NoMachine ensure low bandwidth transmission between the client and the server. Indeed, only user interaction (mouse, keyboard) and local updates of the display are exchanged. The extension of thin client protocols on mobile terminals is currently proposed by several companies such as NEC and HP. The main limitation of the traditional thin client approach is in the poor support of very dynamic content such as movies or games, since the screen updates are transmitted as (sometimes compressed) bitmaps . A direct extension is to perform remote-rendering and transmit the screen as a video stream. With games, this is equivalent to interpreting the graphics primitives on the server (much more powerful than the mobile phone), rendering in a memory buffer, encoding it as a video sequence, and streaming the result to the mobile device [1]. In the upper channel, the user interaction with the terminal (keyboard and screen touch) is directly transmitted to the server. The only processing requirement on the client side is to have a terminal able to decode and display the video stream.

The major limitation of this solution is the need for significant bandwidth, making it feasible only when the network connection can insure good transmission quality. On the server side, this approach requires significant computational power, and leads to poor performances if more users are using the system simultaneously. Additionally, the graphic card of the server has to be shared among the users. As an alternative, the rendering can be distributed on dedicated rendering clusters, as proposed in the Chromium project [2], thus supporting more clients.

In the current chapter, we propose a "thin client" architecture and a protocol dedicated to the efficient handling of 3D graphics games on mobile phones. As with the classical approach, the architecture consists of a client and a server. In opposition, the updates are operated on top of a scene graph (shared by both client and server) and not on the resulting bitmap. Thus, the client is including some intelligence to generate (render) the screen from the scene graph, the prize consisting of a very narrow transmission bandwidth even for very dynamic scenes. We are addressing the problem of terminal heterogeneity by analyzing the 3D graphics capabilities of an on-going multimedia standard, namely MPEG-4 [3], as detailed in Section 2. Section 3 introduces the main elements of the proposed client-server architecture and Section 4 presents the experimental results of a game implemented by following the proposed approach.

2   MPEG-4 3D Graphics for Mobile Phones

Starting from the first version of the MPEG-4 standard, published in 1998, graphics (2D and 3D) was considered within the MPEG community as a new media, with interest comparable to the more traditional video and audio. The objective was to standardize the bitstream syntax able to represent scene graph and graphics primitives, including geometry, texture and animation. These specifications are published in different parts of MPEG-4: Part 11 – BIFS (Binary Format for Scene) treats binarization and compression of a scene graph based on VRML; Part 16 – AFX (Animation Framework eXtension) offers a toolbox for compressing graphics primitives; and Part 25 – 3DGCM (3D Graphics Compression Model) introduces a model for binarization and compression of arbitrary XML scene graph. BIFS includes mechanisms for updating the scene graph, locally or remotely. In [4] and [5] it was demonstrated how the TV provider can use BIFS to include content for commercial advertisements, shopping, electronic on-line voting, on-line quiz program etc. in their program. BIFS is used mainly to transfer graphics data to the client, but also for bidirectional communication.

Using BIFS for streaming 3D content has been explored by Hosseini [6] proposing a powerful JAVA-based framework. However a JAVA-based approach is less appropriate for mobile devices due to decreasing of performances. Using BIFS for games is not completely new, being already exploited by Tran [7] who proposed several 2D games. However, the updates are handled locally, no server being involved. To our knowledge, no other system addressing 3D games with client-server interaction by using BIFS has been published.

Several implementations of MPEG-4 graphics are made available in products such as those proposed by iVast or Envivio, or in open source packages such as GPAC [8]; however, the literature on MPEG-4 3D graphics on mobile phone is almost inexistent. In order to quantify its capabilities for representing graphics assets as used in games, we implemented an MPEG-4 3D graphics player for the Nokia S60 platform based on the Symbian S60 FP1 SDK [9]. The testing hardware included Nokia N93, Nokia N95 and Nokia N95 8GB (which have nearly the same hardware and performances). To ensure good performances, the player was implemented in C and C++. For MPEG-4 de-multiplexing and BIFS decoding, the implementation is full software (based on the GPAC framework). The implementation of the rendering is hardware supported (based on OpenGL ES [10]). One of the objectives of our experiments is to find out the upper limits in terms of 3D assets complexity (with respect to geometry, texture and animation) while ensuring fast decoding (it should be transparent to the player that the asset was first decoded before rendering). The second objective is to analyze the rendering performances of the selected platform.

The first test performed addressed MPEG-4 files containing only static and textured objects with different number of vertices and triangles. We measured the BIFS decoding time and the rendering frame-rate for each file. The second test is related to animated objects based on the skeleton-driven deformation approach. This kind of content, requiring increased computation due to the operation per vertex performed during the animation, leads to lower rendering frame-rate than static objects.

Figures 1 and 2 illustrate a classical behavior of decoding and rendering capabilities against the number of low-level primitives (vertices, normals, triangles ...) for static and animated objects, respectively. The vertical axis on the left represents the BIFS decoding time in milliseconds, and the vertical axis on the right represents the frame-rate in frames per second (FPS). Let us note that considering only vertices on the horizontal axis does not provide a complete analysis for BIFS decoding, since the structure of the object graph is not flat as in the case of the structure used when the object is rendered. Indeed, when defining an object in BIFS it is possible to create different index lists for each primitive (vertex, normals, texture coordinates, colors ...); hence the total number of low-level primitives is reported on the horizontal axis.

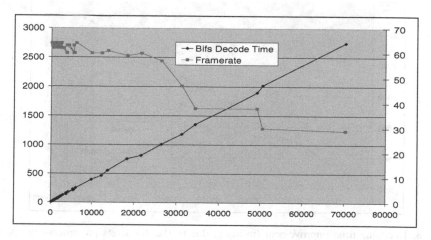

Fig. 1. Decoding time (in ms on left side) and rendering frame-rate (in FPS on right side) for static objects with respect to the total number of low level primitives in the MPEG-4 file

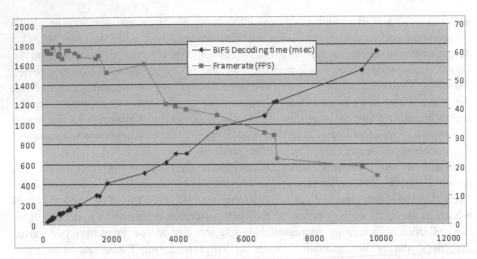

Fig. 2. Decoding time (in ms on left side) and rendering frame-rate (in FPS on right side) for animated objects with respect to the number of triangles in the MPEG-4 file

In order to reduce the decoding time, one solution is to find a more compact manner of representing the low level primitives. We reduce the number of index lists by pre-processing the object in order to flatten its structure: vertices, normals and texture coordinates share a common list. This pre-processing has an impact on the length of each individual primitive's arrays (slightly increasing them), but overall it reduces the size of the compressed data and implicitly the decoding time (around 13%). Figure 3 shows the comparative BIFS decoding time obtained with and without pre-processing.

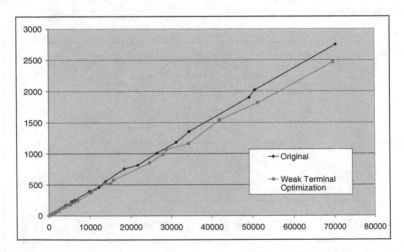

Fig. 3. Decoding time improvement (in msec) due to the 3D assets pre-processing. On the horizontal the total number of low level primitives.

Concerning the capabilities of MPEG-4 for decoding and N95 for rendering, we can observe the following: to obtain acceptable decoding time (less than 2 seconds), a 3D asset represented in MPEG-4 should have less than 58 000 low level primitives (corresponding to objects with less than 15 000 vertices and 12 000 triangles) for static objects and 1 500 triangles for animated objects. N95 is able to render textured and lighted static objects of around 20 000 vertices and 20 000 triangles at an acceptable frame rate (25 fps) and it is able to render at the same frame rate textured and lighted animated objects of around 6 000 vertices and 6 000 triangles. Figure 4 presents snapshots of the player loading different static and animated objects used for tests.

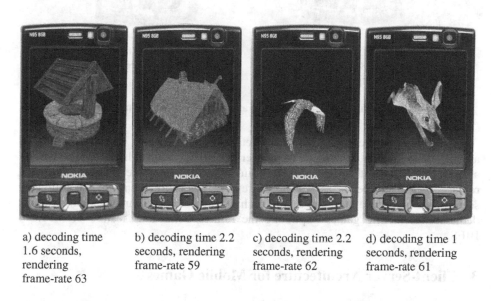

a) decoding time 1.6 seconds, rendering frame-rate 63

b) decoding time 2.2 seconds, rendering frame-rate 59

c) decoding time 2.2 seconds, rendering frame-rate 62

d) decoding time 1 seconds, rendering frame-rate 61

Fig. 4. Snapshots for static (a and b) and animated (c and d) 3D graphics objects

The reduced display size of mobile phones implies that only a small number of pixels are used to render (sometimes dense) meshes. Simplification of geometry and texture can be used without affecting the visual perception of the objects. By doing so, the size of the content can be reduced by 60-80% and the loading time by 70-90% [11]. Figure 5 shows the original Hero model and its version simplified at 27%. The size of the simplified file is 428kb (24% with respect to the original one) and the total loading time is 3.8 sec (57% faster than the original one).

The two tests performed on a large database (around 5 000 graphics files of different nature) indicate that MPEG-4 may offer appropriate representation solutions for simple static and animated objects, both in terms of loading time and rendering frame-rate, by offering a good compromise between compression performances and complexity.

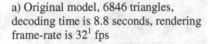

a) Original model, 6846 triangles, decoding time is 8.8 seconds, rendering frame-rate is 32[1] fps

b) Simplified model, 1874 triangles, decoding time is 3.8 seconds, rendering frame-rate is 58 fps

Fig. 5. The original Hero and its simplified version

In addition to representing graphics assets and scenes locally, MPEG-4 introduced a mechanism to update the scene from a server. BIFS-Commands are used to modify a set of properties of the scene at a given time, making it possible to insert, delete and replace nodes and fields, as well as to replace the entire scene. In the following section we introduce a client-server architecture and a communication protocol dedicated to mobile games exploiting the remote commands feature of MPEG-4 BIFS.

3   Client-Server Architecture for Mobile Games

Two main categories of requirements have driven our developments in proposing the client-server architecture:

- for game creators, the deployment of a game on a large category of terminals should not lead to additional development cost,
- for players, the game experience (mainly measured in game reactivity and loading time) should be similar compared with a game locally installed and executed.

The main idea is to separate the different components presented in a traditional game into those that are executed on the server and those that are executed on the terminal. In a simplified schema such as the one illustrated in Figure 6, one may observe two major high processing components in a game: the game logic engine and the rendering engine. The first receives updates from modules such as data parsers, a user

---

[1] The "hero" model is one of the most complex objects we tested, with the following characteristics: 9 disconnected objects, 3587 vertices, 6846 triangles, 9 textures, 63 bones and 31 animation frames.

interaction manager, artificial intelligence, and network components, and updates the status of an internal scene graph. The second is in charge of synthesizing the images by interpreting the internal scene graph.

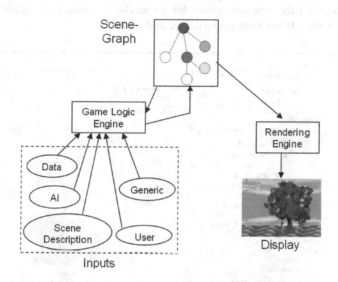

Fig. 6. The main functional components of an arbitrary game

As shown in the previous section, MPEG-4 has the capability of representing (in a compressed form) a scene graph and graphics primitives and an MPEG-4 player is able to interpret them to produce a synthetic image. The main idea proposed here is to replace the rendering engine of the game (right side in Figure 6) with an MPEG-4 player, with the following consequences: during the game, the scene graph (or parts of it) has to be transmitted to the client and the user input (captured by the client) has to be transmitted to the server. Figure 7 illustrates the proposed architecture. The direct advantage is that the MPEG-4 Player is a standard player and does not contain any game specific code.

The main underlying idea of the architecture proposed in Figure 7 is to execute the game logic on the server and the rendering on the terminal. Therefore, different types of games may be accommodated in the proposed architecture. The only strong requirement is the latency allowed by the game play. According to the games' classification with respect to complexity and amount of motion on the screen, as proposed by Claypool [12], and game latency, as proposed by Claypool [13], our goal is to test the architecture and pronounce on its appropriateness and the usage conditions for the four classes: third person isometric games, omnipresent, third person linear games, and first person.

In addition, the player receives only what is necessary at each step of the game (interface 1 in Figure 7). For example, in the initial phase only some 2D graphics primitives representing the menu of the game are transmitted. When the game starts the 3D assets are sent only when they are used, the MPEG-4 compression ensuring fast transmission. During the game-play, the majority of the communication data consists of updates (position, orientation) of assets in the local scene. Let us note that for games containing complex assets it is also possible to download the scene graph, i.e. an MPEG-4 file, before starting playing. The transferring off-line content has similar functionality as the caching mechanism proposed by Eisert [14]. In addition, it is possible to adapt the graphics assets for a specific terminal [15], allowing for the best possible trade-off between performance and quality.

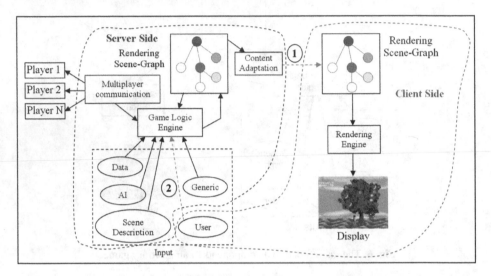

Fig. 7. Proposed architecture for mobile games using an MPEG-4 player on the client side

In the proposed architecture, the communication characterized by interfaces 1 and 2 in Figure 7, unlike in [14], is based on a higher level of control: the graphic primitives can be grouped and controlled as a whole by using fewer parameters. The MPEG-4 standard allows any part of the scene to be loaded, changed, reorganized or deleted. For example, the same game can be played on a rendering engine that supports 3D primitives, most probably accelerated by dedicated hardware, and simultaneously on a rendering engine that only supports 2D primitives. This flexible approach allows the distribution of the games on multiple platforms without the need to change the code of the game logic. Another advantage is the possibility to improve the game logic without additional costs to the client, allowing easy bug-fixing, adding features and different optimizations.

Since the game logic can be hosted on a server that out-performs the client terminal, it can implement more advanced features, traditionally not supported on the client terminal. These features may include advanced artificial intelligence or advanced physics. These improvements will not change anything on the user side, therefore allowing the user to play more complex games on the same terminal.

The flexibility of the proposed architecture makes it also appropriate for powerful terminals with the goal of reusing the rendering engine between different games. Under these circumstances, it is possible to download the game logic server software, install it on the terminal and use it to play the game locally. However, in order to maintain the effectiveness of this approach, the games should be designed from the start bearing in mind the requirements and the restrictions.

The rendering engine (ensured by the MPEG-4 player) can be integrated into the firmware of the device (i.e. supplied with the device), allowing the manufacturer to optimize the decoders and rendering for specific hardware. Let us note that, beside BIFS, which is a generic compression algorithm able to reduce the data size up to 15:1, additional tools exist in MPEG-4 improving the compression of graphics primitives up to 40:1 [16].

The following tables (Table 1 and Table 2) summarize the advantages and disadvantages of the proposed approach with respect to the local game play.

Table 1. Comparative evaluation of the proposed method with respect to a game executed locally

	Local game play	Proposed architecture
Advantages		
Software development	The game should be compiled for each type of terminal	The game logic is compiled once for the dedicated server. The game rendering is ensured by a standard MPEG-4 player, optimized for the specific terminal
Advanced game features	Reduced due to the limitation of the terminal	By choosing high end servers, any modern game feature, such as advanced Artificial Intelligence. can be supported
Rendering frame rate	Since the terminal processing power is shared between game logic and rendering, it is expected that the frame rate is smaller	High because the terminal only performs rendering and asynchronous scene graph decoding
Game state consistency in multi-player games	Synchronization signals should be send between terminals and complex schema should be implemented	Synchronization is directly ensured by the server that controls the scene graph of each terminal at each step
Security	Games can be easily copied	The game starts only after connection to the server, when an authorization protocol may be easily set up
Maintenance and game updates management	Patches should be compiled for each version of the game	Easily and globally performed on the server

Table 2. Comparative evaluation of the proposed method with respect to a game executed locally

Disadvantages		
Network	No impact	The game stops without network connection
Network latency	No impact	The game experience decreases if the loop (user interaction – server processing – update commands – rendering) is not fast enough[2]
Adaptation of existing local games to proposed architecture	No impact	Access to the game source code is recommended, but it is not necessary

It is evident from Table 2 that the main drawback of the proposed method is the sensitivity to network latency. Another weak point is the adaptation of existing games. Two components need to be considered: sending the 3D content to the client and getting the user commands from the client. There are three possibilities for conveying the content to the client:

- If the source code of the game is accessible, then it can be relatively easily modified to fit with the proposed architecture.
- If a high level API is available for the game, supporting access to the graphics objects and the scene-graph, then a separate application can be created that will connect the architecture and the game.
- If no game API is available, then a wrapper for the 3D library that is used by the game can be created. It will extract the geometry and transformations from the graphics calls and it is necessary to implement a converter between graphics commands and BIFS updates.

There are two possibilities for receiving the user commands:

- If the source code of the game is available, it can be modified easily for receiving the commands.
- Otherwise, a software module can be created that will receive the commands from the client, create the appropriate system messages and send them to the game.

To quantify the impact of network latency, we implemented different games and tested them in different conditions. We also defined a methodology in 6 steps to adapt an existing game to a form able to run in the proposed architecture. In the next section, we present a car race game and the experiments we conducted to validate the architecture.

---

[2] The notion of "fast" is game specific. For first person shooter games, the loop performed in 100 ms is acceptable, for sports games, 500ms, and for strategy games 1000ms [5].

4 Implementation, Simulation and Results

Based on the architecture proposed in the previous section, we implemented a multi-player car race. The game state changes frequently, making it appropriate for testing the architecture. The game was originally developed in J2ME as a traditional multiplayer game for mobile phones. It uses a GASP server [17] for communication between the players. Originally, the logic and the rendering engine were implemented in the J2ME software, and the GASP server was used only to transfer the positions between the players. A simple 2D rendering engine rendered the track and the cars as sprites [18].

The adaptation of the game was performed in several steps:

(1) Identifying of the structures used to hold the relevant data (position of the cars, rotation of the camera in the case of the car race).

(2) Identification to the main loop and discarding the rendering calls.

(3) Adding the network communication components in the main loop, both from the game to the player and vice-versa.

(4) Defining the BIFS command messages, encode message encapsulation and message parsing (for sending the position of the cars to the player and receiving the command keys from the player, respectively).

(5) Converting all the assets and the scene graph in MPEG-4 files.

The MPEG-4 player presented in Section 2 was enriched with a BIFS communication layer to ensure connection to the game server. It receives the BIFS-commands and updates the scene accordingly, and it transmits the key pressed by the player to the game server. The key-presses are detected by using a BIFS node called InputSensor. When this node is activated, JavaScript [19] code stored in the scene is executed. The code makes an HTTP request by using AJAX [20] which transfers the pressed key code to the server.

Independent from the game, a communication protocol compliant with the proposed architecture has three phases: initialization, assets transmission and playing. In the following text we define the three phases for the car race game in two configurations, single and multi-user by analyzing at each step the transmitted data.

1) Initialization:

- the player initiates a game by pointing to an *url* referring to a remote MPEG-4 file;
- when a request is received, the server initiates a TCP session and sends the initial scene containing a simple 2D scene presenting a menu with different options: "New Game", "Connect", "Exit" as illustrated in Figure 8a; (the data transmitted is about 0.4 KB)
- the scene also contains an InputSensor and JavaScript which is used to forward the code of a key pressed by the user to the server. The server updates its own state of the scene graph, and if the update has impact on the rendering (e.g. change of the color when an item of the menu is selected), the corresponding BIFS-commands is transmitted to the client (about 0.1 KB).

2) Assets transmission:
- if "New Game" is selected,
  a. the server sends an update command for displaying the snapshots (still images) of several cars in a mosaic (about 40 KB for 4 cars)
  b. when the user selects one snapshot, the server sends a scene update containing the 3D representation of the car (two versions are currently available, 82 KB and respectively 422 KB). The car is received by the player and rendered as an 3D object as illustrated in Figure 8b (the local 3D camera is also enabled);
  c. when the car is validated, the server sends a mosaic of still images with available circuits and the same scenario as for the car is implemented (about 40 KB for the mosaic of 4 tracks, 208 KB and 1.6 MB for the 3D circuits);
- If "Connect" is selected, the server checks for existent sessions (previously initiated by other users) and sends a list;
  a. after selecting one session (implicitly the track is selected), the server sends the mosaic for selecting the car
  b. after selecting the car, the 3D object representing it is transmitted

3) Playing the game
- the code of the key pressed by the user (accelerate or brake) is directly transferred to the server (about 350 bps), that computes the car speed and implements the rules for game logic (points, tire usage, …)
- the server sends the new 3D position of all the cars in the race (about 650 bps), updates for the status icons and the number of points; the player processes the local scene updates as illustrated in Figure 9.

a) Phase 1: Initialization   b) Phase 2: Assets transmission - examples of 3D assets transmission and visualization (car and circuit)

Fig. 8. Snapshot from the car game

Phase 3: Playing the game

Fig. 9. Snapshot from the car game

We set up several experiments to objectively measure the user experience when playing the car race game, based on time to respond to user interaction (phases 1 and 3) and time to wait for transmission and loading of 3D assets (phase 2). Table 3 presents the latency when assets are transmitted and Figure 10 the latency when only user interaction and updates commands are transmitted. The measurements are performed for two network configurations: Wi-Fi (IEEE 802.11, ISO/CEI 8802-11) and UMTS.

Table 3. Latency (transmission and decoding) for the 3D assets used in the car race game

Asset	Car_v1	Car_v2	Circuit_v1	Circuit_v2
Number of vertices	253	552	1286	7243
MPEG-4 file size (KB)	82	422	208	1600
Transmission time Wi-Fi	27	126	68	542
Transmission time UMTS	422	2178	1067	8246
Decoding time (ms)	112	219	328	2538
Total waiting time Wi-Fi (ms)	139	345	396	3080
Total waiting time UMTS (ms)	534	2397	1395	10784

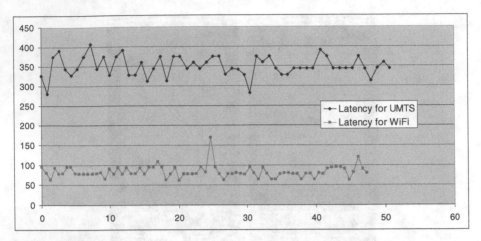

Fig. 10. Time to response (in msec) for UMTS and Wi-Fi recorded during the Phase 3 "Playing the game". The horizontal axis represents the duration of play.

Let us note an average execution time of 80 msec and a maximum of 170 msec for Wi-Fi connection and an average of 350 msec and a maximum of 405 msec for UMTS connection for the entire loop consisting of the transmission of the user interaction (interface 2 in Figure 7), the time for processing on the server, the BIFS-commands transmission (interface 1 in Figure 7) and decoding and rendering of the local scene updates.

To evaluate our results, we have compared them to the research done by Claypool [13] on the effect of latency on users in online games. The paper proposes experiments with different types of games [21, 22, 23, 24] and evaluates how the latency influences the gameplay results. The quality of the gameplay specific for each game is measured. For example, for "first person game" the hit fraction when shooting at a moving target is measured, for driving game the lap time, etc. Then the results are normalized in a range from 0 (worst) to 1 (best) and exponential curves for each game category are obtained. A copy of these curves is represented in Figure 11. The horizontal gray bar – around 0.75 (originally proposed by Claypool [13]) represents the player tolerance threshold with respect to latency. The section above is the area where the latency does not affect the game play performance. The section below is the area where the game cannot be correctly played.

On the same graph, we plot our latency results obtained for the race game. Based on Figure 11 and on the classification of the games according to scene complexity and latency, we can draw some conclusions:

- The architecture is appropriate for omnipresent games for the two network configurations.
- For third-person avatar games, the proposed architecture is appropriate when a Wi-Fi connection is used, and it is at the limit of user tolerance when a UMTS connection is used.
- For first-person avatar games, the proposed architecture is inappropriate when using UMTS connection and it is on the lower boundary of the user tolerance when using Wi-Fi connection.

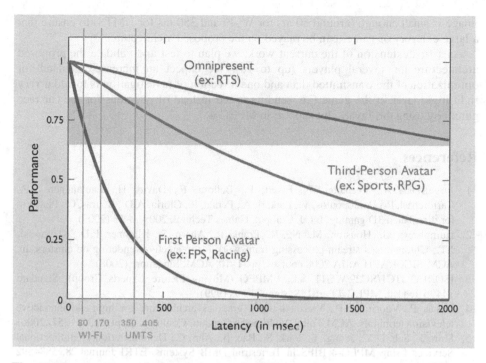

Fig. 11. Player performance versus latency for different game categories. Original image reproduced from [13] with permission.

5 Conclusion

In this chapter we have presented an alternative architecture for games on mobile phones aiming to cope with the problems of costly deployment of games due to the heterogeneity of the mobile terminals. By exploiting and extending some concepts of "thin client" introduced in the early nineties, we based our solution on the 3D graphics capabilities of MPEG-4. We have presented an implementation of an MPEG-4 3D graphics player for mobile phones and evaluated its capabilities in terms of decoding time and rendering performance. Demonstrating the pertinence of using MPEG-4 as a standard solution to represent 3D assets on mobile phones, the main idea of the proposed architecture is to maintain the rendering on the user terminal and to move the game logic to a dedicated server. This approach presents several advantages, such as the ability to play the same game on different platforms without adapting and recompiling the game logic, and to use the same rendering engine to play different games. The use of MPEG-4 scene updates for the communication layer between the game logic and the rendering engine allows the independent development of the two, with the advantage of breaking the constraints on the game performances (by using the server) and of addressing a very fragmented and heterogeneous park of mobile phones. The measurements performed on network and end-user terminals are satisfactory with respect to the quality of the player's gaming experience. In particular, the latency between user interaction on the keyboard and the rendered image is small enough (around 80 ms for Wi-Fi and 350 ms for UMTS) to ensure that a large category of games can be played on the proposed architecture.

As a first extension of the current work, we plan to test and validate the proposed architecture for several players (up to 16). We expect an increased demand for optimization of the transmitted data and one direction of investigation is to add a very light logic layer in the game (i.e. sending speed instead of positions for the car race game) by using the JavaScript support in MPEG-4.

References

1. Jurgelionis, A., Fechteler, P., Eisert, P., Bellotti, F., David, H., Laulajainen, J.P., Carmichael, R., Poulopoulos, V., Laikari, A., Perälä, P., Gloria, A.D., Bouras, C.: Platform for distributed 3D gaming. Int. J. Comput. Games Technol. 2009, 1–15 (2009)
2. Humphreys, G., Houston, M., Ng, R., Frank, R., Ahern, S., Kirchner, P.D., Klosowski, J.T.: Chromium: a stream-processing framework for interactive rendering on clusters. In: ACM SIGGRAPH ASIA 2008 courses, pp. 1–10. ACM, Singapore (2008)
3. ISO/IEC JTC1/SC29/WG11, a.k.a. MPEG (Moving Picture Experts Group): Standard 14496 1, a.k.a. MPEG 4 Part 1: Systems, ISO (1999)
4. Cesar, P., Vuorimaa, P., Vierinen, J.: A graphics architecture for high-end interactive television terminals. ACM Trans. Multimedia Comput. Commun. Appl. 2, 343–357 (2006)
5. Shin, J., Suh, D.Y., Jeong, Y., Park, S., Bae, B., Ahn, C.: Demonstration of Bidirectional Services Using MPEG-4 BIFS in Terrestrial DMB Systems. ETRI Journal 28, 583–592 (2006)
6. Hosseini, M., Georganas, N.D.: MPEG-4 BIFS streaming of large virtual environments and their animation on the web. In: Proceedings of the seventh international conference on 3D Web technology, pp. 19–25. ACM, Tempe (2002)
7. Tran, S.M., Preda, M., Preteux, F.J., Fazekas, K.: Exploring MPEG-4 BIFS features for creating multimedia games. In: Proceedings of the 2003 International Conference on Multimedia and Expo, vol. 2, pp. 429–432. IEEE Computer Society, Los Alamitos (2003)
8. Feuvre, J.L., Concolato, C., Moissinac, J.: GPAC: open source multimedia framework. In: Proceedings of the 15th international conference on Multimedia, pp. 1009–1012. ACM, Augsburg (2007)
9. Edwards, L., Barker, R.: Developing Series 60 Applications: A Guide for Symbian OS C++ Developers. Pearson Higher Education, London (2004)
10. Trevett, N.: Khronos and OpenGL ES, Proceedings of SIGGRAPH 2004, Tokyo, Japan (2004), http://www.khronos.org/opengles/1_X/
11. Preda, M., Villegas, P., Morán, F., Lafruit, G., Berretty, R.: A model for adapting 3D graphics based on scalable coding, real-time simplification and remote rendering. Vis. Comput. 24, 881–888 (2008)
12. Claypool, M.: Motion and scene complexity for streaming video games. In: Proceedings of the 4th International Conference on Foundations of Digital Games, pp. 34–41. ACM, Orlando (2009)
13. Claypool, M., Claypool, K.: Latency and player actions in online games. Commun. ACM 49, 40–45 (2006)
14. Eisert, P., Fechteler, P.: Low delay streaming of computer graphics. In: 15th IEEE International Conference on Image Processing, ICIP 2008, pp. 2704–2707 (2008)

15. Morán, F., Preda, M., Lafruit, G., Villegas, P., Berretty, R.: 3D game content distributed adaptation in heterogeneous environments. EURASIP J. Adv. Signal Process. 2007, 31 (2007)
16. Jovanova, B., Preda, M., Preteux, F.: MPEG-4 Part 25: A Generic Model for 3D Graphics Compression. In: 3DTV Conference: The True Vision - Capture, Transmission and Display of 3D Video, pp. 101–104 (2008)
17. Pellerin, R., Delpiano, F., Duclos, F., Gressier-Soudan, E., Simatic, M., et al.: GASP: An open source gaming service middleware dedicated to multiplayer games for J2ME based mobile phones. In: Proceedings of International Conference on Computer Games, Angoulême, France, pp. 28–30 (2005)
18. Buisson, P., Kozon, M., Raissouni, A., Tep, S., Wang, D., Xu, L.: Jeu multijoueur sur téléphone mobile (in French), Rapport de projet ingénieur, Rapport final projet S4 2007 (TELECOM Bretagne),
    http://proget.int-evry.fr/projects/JEMTU/ConceptReaJeu.html
19. Flanagan, D., David, F.: JavaScript: The Definitive Guide. O'Reilly Media, Sebastopol (2006)
20. Garrett, J.J.: Ajax: A New Approach to Web Applications,
    http://www.adaptivepath.com/ideas/essays/archives/000385.php
21. Beigbeder, T., Coughlan, R., Lusher, C., Plunkett, J., Agu, E., Claypool, M.: The effects of loss and latency on user performance in unreal tournament 2003®. In: Proceedings of 3rd ACM SIGCOMM workshop on Network and system support for games, pp. 144–151. ACM, Portland (2004)
22. Claypool, M.: The effect of latency on user performance in real-time strategy games. Comput. Netw. 49, 52–70 (2005)
23. Nichols, J., Claypool, M.: The effects of latency on online madden NFL football. In: Proceedings of the 14th international workshop on Network and operating systems support for digital audio and video, pp. 146–151. ACM, Cork (2004)
24. Sheldon, N., Girard, E., Borg, S., Claypool, M., Agu, E.: The effect of latency on user performance in Warcraft III. In: Proceedings of the 2nd workshop on Network and system support for games, pp. 3–14. ACM, Redwood City (2003)

A Face Cartoon Producer for Digital Content Service

Yuehu Liu, Yuanqi Su, Yu Shao, Zhengwang Wu, and Yang Yang

Institute of Artificial Intelligence and Robotics,
Xi'an Jiaotong University, Xi'an, P.R. China, 710049
liuyh@mail.xjtu.edu.cn

Abstract. Reproducing face cartoon has the potential to be an important digital content service which could be widely used in mobile communication. Here, a face cartoon producer named "NatureFace", integrated with some novel techniques, is introduced. To generate a face cartoon for a given face involves proper modeling for face, and efficient representation and rendering for cartoon. For both face and cartoon, new definition are introduced for modeling. For face modeling, it is the reference shape; while for cartoons, the painting pattern for corresponding cartoon. Sufficient samples are collected for two parts from the materials supplied by invited artists according to the pre-assigned artistic styles. Given an input face image, the process for generating cartoon has the following steps. First shape features are extracted. Then, painting entities for facial components are selected and deformed to fit current face. Finally, the cartoon rendering engine synthesizes painting entities with rendering rules originated from designated artistic style, resulting in the cartoon. To validate the proposed algorithm, cartoons generated for a group of face images with rare interaction are evaluated and ranked in an artistic style way.

Keywords: Face image, face cartoon, reference shape, painting pattern, rendering engine, digital content service, mobile communication.

1 Introduction and Related Work

Inspired by the desire for stylized communication, some content services show their great potential in application market like the MSN cartoon. Cartoons are widely used in daily life for transferring information, which show themselves in comic books, TV shows, IMs and so on.

In this chapter, we introduce a cartoon producer comprising several artistic styles, helping mobile subscribers to produce and share their own cartoons in a easy way with little input. Given a photo of face, the producer can generate the face cartoons of assigned style such as sketch or color cartoon, and can animate cartoons with interesting facial expressions. Subscribers can then share these cartoons among friends in the form of an MMS (multimedia messaging service).

Before, some impressive works have been developed to produce face cartoon from a given photo, such as PicToon, MSN cartoon, PICASSO and so on.

X. Jiang, M.Y. Ma, and C.W. Chen (Eds.): WMMP 2008, LNCS 5960, pp. 188–202, 2010.

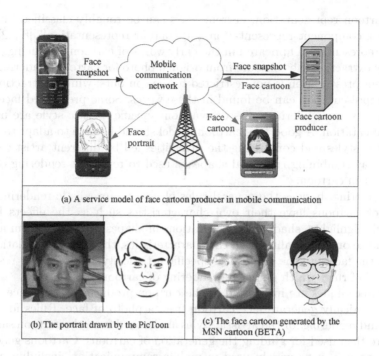

(a) A service model of face cartoon producer in mobile communication

(b) The portrait drawn by the PicToon

(c) The face cartoon generated by the MSN cartoon (BETA)

Fig. 1. (a) gives the framework of cartoon producer used in mobile environment. After a photo of face is taken by cell phone, it is then sent to a cartoon producer through mobile communication network. Then the generated cartoon can be sent back to the subscriber or shared with someone else. (b) and (c) give two influential face cartoon generation systems. (b) is a cartoon producer named "Pictoon", which generates sketches with statistical model. (c) is MSN cartoon which is a online software published by Microsoft.

PicToon utilizes a statistic model to create shapes of cartoon from given face which is subsequently rendered with a non-parametric sampling method. It can generate the facial sketches and animations from templates [3][4], with an example shown in Fig. 1(b). MSN cartoon[5] is an online application published by Microsoft. Human interactions are required by the system to select the cartoon template, adjust the shape profile and add accessories. Fig. 1(c) shows a generated result. The PICASSO system[6] deformed the input face, and generate the facial caricature by comparing the mean face.

It is concluded from previous work, that main issues involved in cartoon producing from a given photo are facial feature extraction, cartoon representation and rendering. For each part, various algorithms have been proposed. However, for the mobile environment, specific requirements should be taken into consideration, including the sensing conditions, image quality and so on.

For facial feature extraction, ASM[1] and AAM[2] are the widely used algorithms. For satisfying the mobile requirements, a modified ASM is used which extracts contours of facial components represented by B-spline with a post-correction process.

For cartoon representation, current ways can be roughly classified into two categories: component representation and contour representation[7][8][12]. The contour representation appeared in the early works of cartoon producing, which faithfully extracting the contours from original photo for cartoon rendering. The component presentation is not restricted to the contours; while abstraction and artistic representation can be found in these works. Some predefined face components, or some rendering rules drawn from specific artistic style are used in the representation. A good face cartoon model should be able to adapt to different artistic styles and comprising the diversities led by different artistic styles. Thus, a way combining rules and shapes is used to represent rendering of each component in cartoon.

For rendering, besides the general rules of non-photorealistic rendering, renderings of cartoons have their own characteristics such as the effects led by reflection, highlights, shadows, exaggerations in shapes originated from artistic styles and so on. For dealing with the characteristics, the statistical relationship between the face and the cartoon of specific artistic style is explored, resulting in a group of rules which guide the rendering of cartoon.

These techniques are integrated in the cartoon producer: NatureFace, which can automatically generate face cartoon from a photo of face. Rules for representing and rendering for the artistic styles are learned from the cartoon samples, which are then used for guiding the generation of cartoons. Cartoons generated by our system can be widely used in mobile communication, including mobile phone entertainment, man-machine interfaces, games and unique idea-sharing. Fig. 1(a) illustrates a service integrating our cartoon producer in mobile communication.

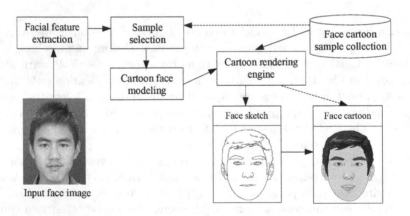

Fig. 2. The framework of the NatureFace system

2   NatureFace: A Face Cartoon Producer

Framework for NatureFace is shown in Fig. 2. There are two processes involved: the offline sample collection and the online cartoon generation.

The sample selection. To generate face cartoon with a given artistic style, a group of samples with same styles are collected, which are further parameterized [18], resulting in the face cartoon sample collection. Cartoon sample collection supplies the generation process with sufficient information, including rendering for each component and the spatial location information for renderings. For gathering the samples, a group of typical face images are picked out and corresponding cartoons drawn by several artistic styles are collected.

The facial feature extraction. The module aims to analyze the face image, and grab the characteristics of a given face. A modified ASM is used here which uses the detection result of face for initialization.

The cartoon face modeling. This module tries to gather the rendering entities, rules and spatial arrangement from the cartoon samples. The cartoon face modeling module calculates the shape parameters of the input face, deforms the selected cartoon component samples and rearranges their location on the face cartoon, creates the cartoon face model for the input face[10].

The cartoon rendering engine. The engine combines the painting entities with the picked out rendering rules and spatial arrangements of the cartoon face model, and outputs face cartoon with artistic style in a vectorized way.

2.1   The Cartoon Face Model

Cartoon face model seeks to give a whole prospect of face and corresponding cartoon. For face, the information is recorded in the reference shape; for cartoon, the painting pattern.

The reference shape defines the facial shape features of a face, denoted as $R = (S, C)$. The facial structure S records the spatial parameters of facial components, while C shapes for each component, which are all represented with B-splines. The used facial components include eyes, eyebrows, nose, mouth, and facial-form.

The painting pattern, denoted as $P = (T, E)$, defines the painting style and the painting entity set for the corresponding face cartoon. T describes the painting style, which is the layout of the painting entities on the face cartoon. In a face cartoon, the painting entity set E includes seven painting entities, one for each used facial component as shown in Fig. 3(c), which are used to synthesize the face cartoon with given artistic style. Each painting entity has triple components: the rendering shape, the rendering elements and the rendering rules.

As mentioned above, The cartoon face model can be defined formally as in Equation (1)-(6).

$$CFmodel = (R, P) \qquad (1)$$
$$R = (S, C) \qquad (2)$$

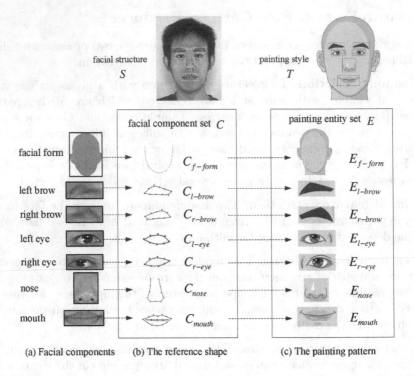

Fig. 3. The proposed cartoon face model. (a) A human face image is decomposed into 7 facial components. (b) The reference shape describes the facial structure and the shape of the facial component, their shape features are approximated with simple geometric shapes. (c) The painting pattern defines the corresponding face cartoon with artistic style using the painting style and the painting entity set. The painting entity consists of the rendering shape, the rendering elements and their rendering rules.

$$C = \{C_{cp}|_{cp \in \{f-form,l-brow,r-brow,l-eye,r-eye,nose,mouth\}}\} \tag{3}$$

$$P = (T, E) \tag{4}$$

$$E = \{E_{cp}|_{cp \in \{f-form,l-brow,r-brow,l-eye,r-eye,nose,mouth\}}\} \tag{5}$$

$$E_{cp} = (E_{cp}^{shape}, E_{cp}^{element}, E_{cp}^{rule}) \tag{6}$$

In the model, rendering element is divided into the curve element and region element, each attached with corresponding rendering rule. Rendering rule specifies the property of the rendering element, decides its appearance. Common properties for the two types contain: layer, visibility, rendering order. The private properties for the curve element include color, degree, weight, line style, brush type, and brush Z-value; for the region element, color, boundary, and filling style. These properties make up the rendering rules, determine the rendering method of the rendering element.

An instance of the cartoon face model can be found in Fig. 3.

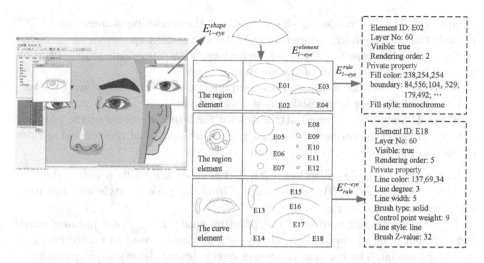

Fig. 4. The FaceEdit system that we develop for parameterizing the face cartoon sample(left). The painting entity E^s_{l-eye} of the parameterized left eye component is consist of the rendering shape E^{shape}_{l-eye}, 13 region elements, 5 curve elements and the corresponding rendering rules. The rendering rules of element E02 and E18 is given as an instance(right).

2.2 The Face Cartoon Sample Collection

Collecting samples from artists just constitutes the first step in building the face cartoon sample collection. Each sample collected is further parameterized[18], and represented as a parameter pair (R^s, P^s): the reference shape parameter R^s of the face, and the painting pattern parameter P^s of the face cartoon.

For parameterizing samples, four steps are involved with the aid of developed parametric tool named "FaceEdit".

Step 1) Extract the landmarks for the contour of facial components, denoted as $F_{contour} = \{x_1, y_1, \cdots, x_{80}, y_{80}\}$, those for painting entity contour of the corresponding cartoon as $C_{contour} = \{u_1, v_1, \cdots, u_{80}, v_{80}\}$ with the Active Contour Model method automatically[1][9].

$F_{contour}$ are used to record the shape and location of each facial component on the face; determine the facial structure S^s and component shape C^s.

$$R^s = (S^s, C^s) \qquad (7)$$

$$C^s = \{C^s_{cp}|_{cp \in \{f-form, l-brow, r-brow, l-eye, r-eye, nose, mouth\}}\} \qquad (8)$$

In the same way, the painting style T^s and the rendering shape of the painting entity set E^s can also be calculated using the contour points $C_{contour}$.

$$P^s = (T^s, E^s) \qquad (9)$$

$$E^s = \{E^s_{cp}|_{cp \in \{f-form, l-brow, r-brow, l-eye, r-eye, nose, mouth\}}\} \qquad (10)$$

$$E^s_{cp} = (E^{shape}_{cp}, E^{element}_{cp}, E^{rule}_{cp}) \qquad (11)$$

Where, E_{cp}^{shape} is the rendering shape, $E_{cp}^{element}$ the rendering element, and E_{cp}^{rule} records the corresponding rendering rules for component cp.

Step 2) Calculate the shape of mean-face, with Euclidean transformation allowed to align the shapes of the samples.

Step 3) Decompose a sample into components, with each component permitted to be selected for synthesizing new face cartoon.

Step 4) Define the layered attributes for rendering elements of the painting entity, set the common properties and private properties of the rendering rules of the painting entity with the cartoon artistic style.

A face cartoon sample collection, with 280 member are built, which is denoted as $\{(R_i^s, P_i^s), 1 \leq i \leq 280\}$. Some samples from the realistic style and beautiful style are illustrated in Fig. 5.

Fig. 4 gives an illustration of the painting entity E_{l-eye}^s of a parameterized left eye. For a facial component, these samples can be used to create new cartoon face model. The cartoon rendering engine learns these sample parameters automatically to draw and synthesize a new face cartoon with given artistic style.

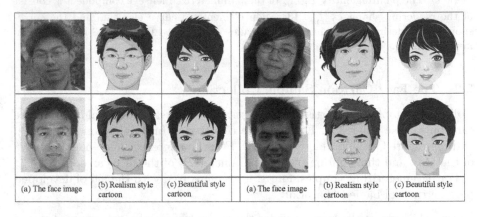

| (a) The face image | (b) Realism style cartoon | (c) Beautiful style cartoon | (a) The face image | (b) Realism style cartoon | (c) Beautiful style cartoon |

Fig. 5. A group of samples drawn by the human artist with realistic-looking style and beautification style

3   Creating the Cartoon Face Model

Given an input photo I^{face}, (R^{face}, P^{face}) denotes its cartoon face model. The goal for creating a cartoon face model is to determine the reference shape $R^{face} = (S^{face}, C^{face})$ and the painting pattern $P^{face} = (T^{face}, E^{face})$.

The reference shape R^{face} can be determined with modified ASM algorithm, which extracts the facial contour points of the input face. These contour points can then be used to calculate the facial structure S^{face} of the input face, and shape C^{face} for each facial component. Due to the impacts of light conditions and facial gestures, the extracted facial contour features can not always be accurate, thus a correction process is adopted to reduce the deviation of extracted shape based on the priori knowledge about facial component distribution[11][13].

For painting pattern P^{face}, its painting style T^{face} is determined by learning statistical relationship from the face cartoon samples. And the painting entity set E^{face} are achieved through selection of proper sample via the best matching and subsequent deformation accordingly.

3.1   Computing the Painting Style

Suppose the location of a facial component is $L = (x, y, w, h, \theta)$. The location of the corresponding painting entity in the painting style is denoted as $L_p = (x_p, y_p, w_p, h_p, \theta_p)$, which is determined by using initial location estimation and sequential optimization. Details are shown in Fig. 6.

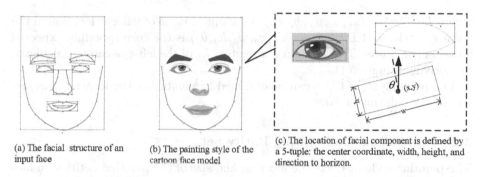

(a) The facial structure of an input face

(b) The painting style of the cartoon face model

(c) The location of facial component is defined by a 5-tuple: the center coordinate, width, height, and direction to horizon.

Fig. 6. The location of the facial component and the corresponding painting entity, which is defined by 5-tuple

For a painting entity, the initialization of its location is as follows:

(1) Set an initial value $L_p^{estimate}$ by using the location of facial component of the input face.

(2) Assume that facial structure is the liner combination of those in the sample collection, and $S = \sum_{k=1}^{N} \alpha_k S_k^s$. The coefficients $\{\alpha_k, k = 1, \cdots, N\}$ are the weights for the samples, which are constrained to $\alpha_k \geq 0$, $\sum_{k=1}^{N} \alpha_k = 1$. Using these coefficients and the painting style $\{T_k^s, k = 1, \cdots, N\}$ of the samples, the expected location value of the painting entity can be obtained.

$$L_p^{optimal} = \sum_{k=1}^{N} \alpha_k T_k^s \tag{12}$$

The initial value of each painting entity and its expected location are sequentially optimized to compose the painting style parameters of the input face. The eyes are optimized firstly, and the location of eyebrows, nose and mouth are determined following optimized eyes. The optimization process makes the painting style approximate to the specific face structure and keep the relative location of the facial component.

For example, the optimization for the left-eye painting entity is represented as the following cost function:

$$\min(\lambda_1 E_w + \lambda_2 E_h + \lambda_3 E_x + \lambda_4 E_y + \lambda_5 E_\theta) \tag{13}$$

$$E_w = (1 - \lambda_w)(w - w_e)^2 + \lambda_w(w - w_o)^2 \tag{14}$$

$$E_h = (1 - \lambda_h)(h - h_e)^2 + \lambda_h(h - h_o)^2 \tag{15}$$

$$E_x = (1 - \lambda_x)(x - x_e)^2 + \lambda_x(x - x_o)^2 \tag{16}$$

$$E_y = (1 - \lambda_y)(y - y_e)^2 + \lambda_y(y - y_o)^2 \tag{17}$$

$$E_\theta = (1 - \lambda_\theta)(\theta - \theta_e)^2 + \lambda_\theta(\theta - \theta_o)^2 \tag{18}$$

where $L^{estimate} = (x_e, y_e, w_e, h_e, \theta_e)$ is the initial value of left-eye location in the painting style, and $L^{optimal} = (x_o, y_o, w_o, h_o, \theta_o)$ is the corresponding expected location parameter. The location parameters L_p of the left-eye can be calculated by deriving Equation (13).

The painting style T^{face} can be obtained by combining the location parameters of the painting entities.

3.2   Computing the Painting Entity Set

The painting style restricts the location and size of the painting entities on face cartoon. After the painting entity set is achieved, only Euclidean transform is allowed to be used to deform each painting entity, with rendering rules replicated from the selected component sample.

The painting entity set E^{face} corresponding with a facial component C_{cp}^{face} can be calculated by Equation (19), where ϕ denotes the conversion function.

$$E^{face} = \{\phi(C_{cp}^{face})|_{cp \in \{f-form, l-brow, r-brow, l-eye, r-eye, nose, mouth\}}\} \tag{19}$$

Suppose that the shape of facial component is represented as $F_{contour} = \{F_i = (x_i, y_i)|i = 1, \cdots, N\}$, and corresponding rendering shape of the painting entity is $C_{contour} = \{C_i = (u_i, v_i)|i = 1, \cdots, N\}$, with i loop for the facial component. There is a one-to-one correspondence between points. TPS transform for non-rigid mapping is used $f = (fu, fv)$ such that the energy defined in Equation (20) is minimized[16][17].

$$E(f) = \frac{1}{n} \sum_{i=1}^{n} \|F_i - f(C_i)\|^2 + \lambda \int \int (f_{xx}^2 + 2f_{xy}^2 + f_{yy}^2)dxdy \tag{20}$$

Here, λ is a parameter balancing the smoothness of f and the deviation of point matching. For a fixed λ, the form for f is as in Equation (21).

$$fu(x, y) = a_{10} + a_{11}x + a_{12}y + \sum_{i=1}^{n} w_{1i}U(\|(x, y)^T - C_i\|)$$

$$fv(x, y) = a_{20} + a_{21}x + a_{22}y + \sum_{i=1}^{n} w_{2i}U(\|(x, y)^T - C_i\|) \tag{21}$$

Here, $U(r) = r^2 lnr$; the coefficient $a_{10}, a_{11}, a_{12}, a_{20}, a_{21}, a_{22}, w_{1i}$, and w_{2i} can be solved by substituting Equation (21) into Equation (20). The selected painting entity is then deformed by f for enforcing the rendering shapes of the painting entity to fit the corresponding shapes of the facial component.

4   The Face Cartoon Rendering

For generating the face cartoon of an input face,there are three steps. The first step is to create the cartoon face model; second to composite the painting entity layer by layer, loyal to the rendering rules corresponded each rendering elements in the painting entity; last, to explain the generated entities according to the painting style of the cartoon face model in image format, and generate a personalized face cartoon with artistic style. Face cartoon rendering is responsible for the last step.

As illustrated in Sect. 2.2, the painting entity consists of two kinds of basic rendering elements: the curve element and the region element. By combination of various basic elements, different artistic styles can be expressed. Each rendering algorithm serves for each rendering element.

Rendering curve element includes generating path and curve rendering. Generating path determines the shape and rendering parameters to be rendered such as the brush type, weight, anti-aliasing parameters and so on. Rendering curve is expressed by the z-buffer brush [14][15], with different brush weight changed for different curves with different style.

The region element can be rendered by the rendering rules, which include the region boundary, fill color and fill style, several algorithms can be found[16].

Using the cartoon rendering engine, the unified approach can be used to render the painting entity by learning the rendering rules, and synthesize the personalized face cartoon with different artistic styles.

5   Experiment Analysis and Evaluation

In order to verify the effectiveness of the proposed algorithm, two types of experiments are conducted: measuring the shape feature error and generating face cartoons with all supplied artistic styles for the input face images.

For an input face, its contour of ground truth is manual aligned with the facial components, denoted as $R = \{r_i | i = 1, 2, \ldots, n\}$. At same time, the rendering shape of the corresponding face cartoon are calculated by the cartoon face modeling approach, the performance of face cartoon generation is influenced by the rendering shape. Fig. 7 illustrates the actual meaning of each contour and rendering shape.

Suppose $F = \{f_i | i = 1, 2, \ldots, n\}$ denotes the rendering shape of painting entity of a cartoon face model for the photo of face, two measurements, the absolute error and the relative error, are used to quantify their shape feature deviation between the rendering shape F and the ground truth R for each facial component.

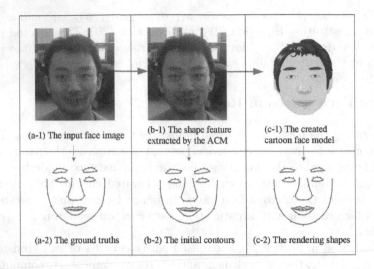

Fig. 7. The ground truth of an input face, initial contours extracted by ACM algorithm and rendering shapes in corresponding cartoon face model

The absolute error e and the relative error χ are measured with formula (22), where $\bar{r}$ is the mean, σ the mean square error.

$$
\begin{cases}
\bar{r} = \frac{1}{n} \sum_{i=1}^{n} r_i \\
\sigma = \sqrt{\frac{1}{n} \sum_{i=1}^{n} (r_i - \bar{r})^2} \\
e = \sqrt{\frac{1}{n} \sum_{i=1}^{n} (f_i - r_i)^2} \\
\chi = \dfrac{e}{\sigma}
\end{cases}
\tag{22}
$$

Experiment 1: We randomly select 49 representative female face images and 49 male face images, 50 of them are frontal face images, 23 are less than 15 degrees to left or right and 25 are less than 20 degrees. These face images are normalized in same size 512×512, resulting in a small test face database shown in Fig. 8. The corresponding ground truthes are manually labeled.

According to the cartoon face modeling approach from an input face image, the initial contours of facial components are firstly extracted by the ACM algorithm automatically, and the rendering shapes of corresponding cartoon face model are then calculated. The changes of shape feature error between the face cartoon and the corresponding face image are exhibited by comparing the ground truth with the initial contour and the rendering shape respectively.

The average relative error and the average absolute error of facial component including the facial form, left eye, right eye, left eyebrow, right eyebrow, nose and mouth, are calculated respectively. These are used to quantitatively

Fig. 8. The test face database including 98 face images

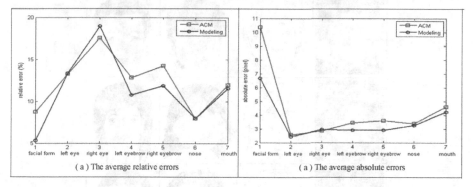

Fig. 9. Comparison of the average relative errors and absolute errors of proposed cartoon face model against the ACM, which are measured with the ground contour using all of faces in database

evaluate the actual effects of the cartoon face modeling for maintaining specific face characteristic.

As can be seen in Table 1, after creating cartoon face model, for each painting entity, the average relative error between rendering shape and ground contour becomes progressively smaller, which indicate that the cartoon face model could keep close to the facial shape features. Particularly, the deviation of the facial-form reduces to nearly its half of the ACM, which indicates the effectiveness of the shape conversion in the painting entity. Table 2 lists the average absolute error, its changing trend is same as the average relative error.

These experimental data indicate that the cartoon face model keeps the basic shape features of the input face and guaranties the usability of the proposed cartoon face modeling.

Table 1. The average relative error(unit:%)

	facial form	left eye	right eye	left eyebrow	right eyebrow	nose	mouth
ACM	8.73	13.27	17.57	12.85	14.23	7.96	11.97
Cartoon face modeling	5.18	12.21	17.86	9.02	9.75	7.49	10.65

Table 2. The average absolute error(unit:pixel)

	facial form	left eye	right eye	left eyebrow	right eyebrow	nose	mouth
ACM	10.35	2.56	2.84	3.44	3.59	3.38	4.58
Cartoon face modeling	6.67	2.43	2.97	2.89	2.91	3.23	4.21

Fig. 10. Generated face sketches and face cartoons with different artistic styles using NatureFace

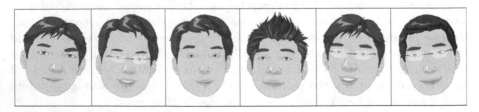

Fig. 11. More generated face cartoons from photo of faces. These input faces have different poses, which are selected from the test face database.

Experiment 2: Generating face cartoons from some input face images using the developed face cartoon producer system. Fig. 10 shows the face cartoon images with different style automatically produced by the NatureFace system. We can see that generated face cartoons keep the original facial features, and have specific artistic style. More generated face cartoons are illustrated as Fig. 11, their input face images are selected from the test face image, these face are not frontal. The NatureFace can maintain their pose in generated cartoons as you can see.

In fact, hairstyle[3] plays an important role in cartoon which makes generated face cartoon more close to the cartoons in reality. In proposed face cartoon producer, the hairstyle is selected from predefined hairstyle sample database, which can not always meet the requirements of cartoon generation for specific input face, especially female face. So appropriate cartoon hairstyle composition is necessary, but which is obviously very challenging work.

6   Conclusions and Further Works

In this chapter, we focus on building a face cartoon producer, named "NatureFace", which can generate the personalized face cartoon from a given face image without rare interaction. As a funny digital media content service, the face cartoon could be widely used in mobile communication.

A novel cartoon face model is proposed, which includes the reference shape and the painting pattern, reference shape defines the shape feature of the face, and painting pattern records rendering of the corresponding cartoon. Some face images and the corresponding cartoons drawn by the artist make up the face cartoon sample collection, each sample is parameterized with the help of the parametric tool "FaceEdit", the samples include the rendering rules and the statistical relationship between the face and the cartoon. For an input face image, the suitable component samples are selected from the face cartoon sample collection, which are adjusted to fit the facial component, and create the cartoon face model by learning the relationship and rendering of the selected samples. The developed rendering engine can synthesize the face cartoon with the artistic style based on the cartoon face model.

The two experiments verify the effectiveness of proposed cartoon model and main technologies. In fact, the developed NatureFace system have be practical applied in mobile communication, as a face cartoon producer, which generates the personalized face cartoon and sends it in the form of an MMS.

Future work will focus on face cartoon generation of different gesture, face exaggeration, and cartoon rendering from 3D face model.

Acknowledgment. This work was supported by the NSF of China (No. 60775017), and the Research Fund for the Doctoral Program of Higher Education of China (No.20060698025). The authors would like to thank Ying Zhang and Shenping Ou for contributing to the face cartoon samples, Daitao Jia and Yuanchun Wang for completing source codes of the cartoon producer. The alignment and analysis of test face images are provided by Ping Wei. We also would like to thank Praveen Srinivasan of University of Pennsylvania for his valuable suggestion and the help of improving the readability of this chapter.

References

1. Blake, A., Isard, M.: Active Contours. Springer, Heidelberg (1998)
2. Cootes, T.F., Taylor, C.J.: Constrained Active Appearance Model. In: IEEE Int'l Conf. Computer Vision (2001)
3. Chen, H., Liu, Z., Rose, C., Xu, Y., Shum, H.Y., Salesin, D.: Example-Based Composite Sketching of Human Portraits. In: The 3rd Int'l symposium on Non-photorealistic aniamtion and rendering, pp. 95–101 (2004)
4. Chen, H., Zheng, N., Liang, L., Li, Y., Xu, Y., Shum, H.Y.: A personalized Image-Based Cartoon System. Journal of Software 13(9), 1813–1822 (2002)
5. MSN cartoon, http://cartoon.msn.com.cn
6. Tominaga, H., Fujiwara, M., Murakami, T., et al.: On KANSEI facial image processing for computerized facial caricaturing system PICASSO. In: IEEE International Conference on SMC, VI, pp. 294–299 (1999)
7. Xu, Z., Chen, H., Zhu, S., Luo, J.: A Hierarchical Compositional Model for Face Representation and Sketching. IEEE Transaction on PAMI 30(6), 955–969 (2008)
8. Liu, Y., Su, Y., Yang, Y.: A facial Sketch Animation Generation for Mobile Communication. In: IAPR Conference on Machine Vision Applications: MVA 2007, Japan (2007)
9. Ratsch, G., Onoda, T., Muller, K.R.: Soft Margin for AdaBoost. Machine Learning 42(3), 287–320 (2001)
10. Liu, Y., Zhu, Y., Su, Y., Yuan, Z.: Image Based Active Model Adaptation Method for Face Reconstruction and Sketch Generation. In: The First International Conference on Edutainment 2006, pp. 928–933 (2006)
11. You, Q., Du, S.: The Splicer for facial Component Image. Technique Report, Institute of Artificial Intelligence and Robotics. Xi'an Jiaotong University (2006)
12. Deng, Z., Neumann, U.: Data-Driven 3D Facial Animation. Springer, Heidelberg (2008)
13. Liu, Y., Yang, Y., Shao, Y., Jia, D.: Cartoon Face Generation and Animation. Technique Report, Institute of Artificial Intelligence and Robotics. Xi'an Jiaotong University (2008)
14. Smith, A.R.: "Paint": in Tutorial: Computer Graphics, pp. 501–515. IEEE Press, Los Alamitos (1982)
15. Whitted, T.: Anti-Aliased Line Drawing using Brush Extrusion. Computer Graphics 17(3), 151–156 (1983)
16. Rogers, D.F.: Procedural elements for computer graphics, 2nd edn. Mc Graw Hill Press (1998)
17. Chuia, H., Rangarajan, A.: A new point matching algorithm for non-rigid registration. In: Computer Vision and Image Understanding, vol. 89, pp. 114–141 (2003)
18. Liu, Y., Su, Y., Shao, Y., Jia, D.: A Parameterized Representation for the Cartoon Sample Space. In: Boll, S., Tian, Q., Zhang, L., Zhang, Z., Chen, Y.-P.P. (eds.) MMM 2010. LNCS, vol. 5916, pp. 767–772. Springer, Heidelberg (2010)

Face Detection in Resource Constrained Wireless Systems

Grigorios Tsagkatakis[1] and Andreas Savakis[2]

[1] Center for Imaging Science, [2] Department of Computer Engineering
Rochester Institute of Technology, Rochester New York 14623
gxt6260@rit.edu, andreas.savakis@rit.edu

Abstract. Face detection is one of the most popular areas of computer vision partly due to its many applications such as surveillance, human-computer interaction and biometrics. Recent developments in distributed wireless systems offer new embedded platforms for vision that are characterized by limitations in processing power, memory, bandwidth and available power. Migrating traditional face detection algorithms to this new environment requires taking into consideration these additional constraints. In this chapter, we investigate how image compression, a key processing step in many resource-constrained environments, affects the classification performance of face detection systems. Towards that end, we explore the effects of three well known image compression techniques, namely JPEG, JPEG2000 and SPIHT on face detection based on support vector machines and Adaboost cascade classifiers (Viola-Jones). We also examine the effects of H.264/MPEG-4 AVC video compression on Viola-Jones face detection.

Keywords: Face detection, Image Compression, Embedded Computer Vision, Resource Constrained Processing.

1 Introduction

The emergence of the terms "ubiquitous computing" and "pervasive computing" during the past decade suggests a new paradigm of human-computer interaction, where various computing devices are dispersed throughout the environment and assist in improving the quality of life [1]. Ubiquitous computing involves systems that incorporate sensing and processing of information in compact and inexpensive platforms. Smart cameras are systems that realize these qualities. They can be found in many everyday devices such as mobile phones, personal digital assistants and portable gaming systems, as well as in more sophisticated systems including robotic platforms and wireless video sensor networks.

Smart cameras represent typical examples of resource constrained embedded systems. The limitations in processing and communication resources have a significant impact on the accuracy, delay and image quality. From the designer's point of view, the most important design issues are constraints in terms of processing capabilities, memory, communication bandwidth and power availability. When it

X. Jiang, M.Y. Ma, and C.W. Chen (Eds.): WMMP 2008, LNCS 5960, pp. 203–220, 2010.

comes to applications, smart cameras provide a new hardware platform for computer vision systems with several advantages including reduced size, lower power consumption and reliability [2].

Regardless of specific architectural system design, a high level model for an embedded vision system is presented in Figure 1. We can identify five fundamental building blocks: an imaging sensor, a processing unit, a memory module, and a communications subsystem. The final element, the battery, is assumed to be interconnected to all other components.

The image sensor may be based on either complementary metal-oxide-semiconductor (CMOS) or charge-couple device (CCD) technology. While CCDs offer higher image quality, in CMOS designs the image sensor and the image processing algorithm can be integrated on the same chip making it easier to optimize the performance in terms of power consumption. The central processing unit and the communications module are typically implemented on System-on-Chip (SoC) platforms in order to meet the processing needs of computer vision algorithms. Examples of hardware processing units include field programmable gate arrays (FPGA), digital signal processors (DSP) and microprocessors. Communication protocols are usually application specific. For example, the IEEE 802.15.4 is a network protocol that is employed in wireless video sensor networks, whereas 3G and Bluetooth are widely used in mobile systems.

Despite their limitations, embedded vision systems offer a new realm for computer vision applications including segmentation, recognition and tracking. In this chapter, we focus on face detection, a necessary step in many high level computer vision applications. More specifically, face detection in an embedded system can be used for people detection and counting for surveillance applications, identification and tracking for smart homes, facial expression and understanding for new human-computer interfaces, gesture, posture and gait analysis for medical monitoring and many more [3].

The rest of the chapter is organized as follows. In Section 2, we overview various image compression techniques and discuss how they have been applied in wireless embedded systems. In Section 3, we provide an overview of face detection methods found in literature and consider the migration of these methods to embedded computer vision systems. Face detection under power constrains is discussed in Section 4. Section 5 provides background information on two face detection algorithms that we investigate in this chapter, namely Support Vector Machines and the Viola-Jones method. Experimental results on the effects of image compression on face detection are presented in Section 6. The chapter concludes in Section 7 with a discussion on the experimental results and some guidelines for system designers.

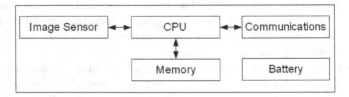

Fig. 1. Block diagram of a sensing node

2 Image Compression Techniques

In the context of power constrained systems, we investigate well-established general purpose image compression algorithms, instead of application specific ones, because of their broader use. More specifically, we investigate face detection in the presence of artifacts introduced by JPEG, JPEG2000 and Set Partitioning in Hierarchical Trees (SPIHT) compression. Most image compression algorithms are based on three steps: transform mapping, quantization and entropy encoding. Compression is achieved by discarding information, such as high frequency components, during quantization followed by effective encoding. The rate of compression and the quality of the reconstructed image are controlled by adjusting the amount of information discarded by quantization.

JPEG, one of the oldest and most successful image compression standards, can achieve good image quality with moderate complexity [4]. In JPEG, an image is first divided into 8x8 non-overlapping blocks, each block is transform-coded using discrete cosine transform (DCT), and the DCT coefficients are ordered in a zigzag order and quantized. The quantization tables used in JPEG are optimized for minimum perceptual distortion. The non-zero coefficients are entropy encoded to generate the final bitstream.

JPEG2000 is a still image compression standard that offers excellent image quality and versatility at the expense of increased complexity [5]. JPEG2000 is wavelet-based and allows lossless to lossy image compression. Initially the image is decomposed into subbands using the wavelet transform and then quantized. Each subband is divided into 64x64 blocks that are independently encoded. Progressive fidelity or progressive resolution schemes are available. For each bit-plane the information related to the most significant coefficients i.e. the ones with the largest magnitude, is encoded first and the bit allocation process continues until all the bit-planes are encoded or the bit budget is met.

The SPIHT compression method is based on wavelet coding [6], and consists of two major components: the set partitioning and the spatial orientation tree. The set partitioning performs an iterative sorting of the wavelet coefficients. At each iteration loop, the significant bits of the coefficients with magnitudes contained in a predetermined region are encoded. The second part, the generation of the spatial orientation trees, is based on the observation that most of the energy is contained in the low levels of the subband pyramid and that there is an inherent self-similarity between subbands.

JPEG is the oldest of the three compression methods considered and offers lower encoding quality compared to JPEG2000 and SPIHT. On the other hand, it is also the most attractive for resource constrained environments because of its low computational complexity. This observation is supported by experimental results reported in [7], where the authors performed an extensive evaluation of various image compression techniques including JPEG, JPEG2000 and SPIHT. Their results for lossy and non-progressive compression showed that JPEG2000 offers the best compression efficiency. SPIHT was a close second and JPEG scored the lowest. The authors also examined the computational complexity of each technique in terms of execution time on a standard PC. They reported that JPEG is the most computationally efficient algorithm with a large margin, compared to JPEG2000 and

SPIHT. SPIHT is in fact the most time consuming encoding algorithm out of these three. The increased computational complexity of JPEG2000 offers a wide variety of functionalities, including random access, region-of-interest encoding, error-resilience, SNR scalability, etc.

Image compression exploits spatial redundancy in the image in order to reduce its size in terms of bits, while minimizing the degradation in quality. Video compression exploits both spatial and temporal redundancy in order to achieve the same goal. A wealth of video compression methods have been developed ranging from the early H.261 coding standard to the H.264/MPEG-4 Advanced Video Codec [39]. Video encoders exploit temporal redundancy by applying motion estimation to predict the location of a region from a previous frame to the current one. Unfortunately, the benefit of reduction in bandwidth is associated with an increase in computational complexity, since motion estimation is a computationally demanding process. As a result, only a few of the current mobile systems employ video encoding.

2.1 Image Compression and Transmission in Mobile Systems

Encoding and transmitting a video sequence is a task that overwhelms most embedded video sensors due to the high complexity of the motion estimation required in video compression protocols based on MPEGx and H.26x series. For this reason, many systems simply encode individual frames using one of the standard image compression techniques. The image/video encoding and transmission systems can be coarsely divided in two categories, the ones that perform encoding individually, such as smart phones, and the ones that use distributed source coding techniques, such as wireless video sensor networks [38].

Although distributed source coding has proven to be very valuable in densely populated wireless video sensor networks, the idea is based on the premise that there is a large overlap between the fields-of-view of various cameras and that the corresponding correlation can be exploited for image/video compression. However, when there is not enough correlation between the sources, there are no significant benefits in using distributed source coding techniques. In this chapter we focus on how encoding of image and video data obtained from an individual camera affects face detection.

In addition to the previous categorization, image compression can be divided into two classes based on the robustness of each method to channel errors. Single layer techniques, as the one employed by JPEG, do not introduce any additional redundancy and thus any additional overheads, but are susceptible to transmission errors. JPEG2000 offers multi-layer coding functionality that can increase the robustness of the method in the presence of transmission errors at the cost of an increase in bitstream size.

In a similar context, Delac et al. [8] provided an overview on experiments that had been reported in the literature regarding the effects of JPEG and JPEG2000 image compression on face recognition. For both classification schemes, the recognition accuracy was only marginally affected in very low encoding rates (below 0.2 bpp), whereas in some cases, a slight increase in the accuracy was also identified.

Due to the importance of image compression and transmission, the relationships between specific image compression techniques and communication protocols have been extensively studied. In [9], the authors compared JPEG and JPEG2000 with respect to transmission over wireless networks based on the IEEE 802.15.4 and ZigBee protocols and noted that JPEG is more suitable for cases where the nodes have to acquire and compress the images because of its low computational complexity, while JPEG2000 is more suitable for networks where packet loss tolerance is required due to frequent bit errors and therefore error resilience is more important.

In [10], the authors considered the energy requirements for transmission of JPEG2000 compressed images over wireless multi-hop networks. They identified a set of three parameters that describe the overall system: transform level, quantization size and communication path loss. Their results suggest that for networks with a small number of hops and high error rates, uncompressed image transmission may achieve higher quality with the same energy consumption compared to compressed ones. The authors also reported that performing entropy coding directly on the raw images can significantly reduce the energy consumption compared to performing JPEG2000 encoding.

In [11], Lee et al. investigated the energy consumption of image compression using JPEG followed by transmission. The authors argue that although JPEG2000 may offer higher compression rates, the full-frame processing requirements are prohibitive for resource constrained embedded systems, as opposed to the 8x8 block processing, employed in JPEG. With respect to compression, they reported a significant decrease in computational time, bitstream size and energy consumption by using a modified version of JPEG compression, optimized for the ATmega128 embedded processor. In terms of transmission, the authors reported that there is a dependence on the compression-transmission pair in terms of the overall energy consumption and bandwidth utilization. In general, there is an advantage in end-to-end processing time and energy consumption when using JPEG compression compared to transmission of uncompressed images in most scenarios investigated.

Most of the work on image/video encoding and transmission in embedded vision systems is primarily focused on how to achieve good image quality, expressed in terms of Mean Square Error (MSE) or peak signal-to-noise ratio (PSNR), while minimizing the power consumption and respecting the limited capabilities of the systems. In this chapter, we investigate image compression with respect to the type of classifier used for face detection. Even though the field of embedded vision is relatively new, there has been some work that examines the migration of successful face detection algorithms in embedded systems, as described in the next section.

3 Overview of Face Detection

A face detection algorithm takes an input image, identifies the human faces and possibly reports their location. Face detection is often a first step needed for other applications including face recognition, face tracking, expression analysis, etc. Image classification algorithms, such as face detection and face recognition, have received considerable attention in the past 20 years. Two observations can be made concerning the nature of the algorithms developed thus far. First, a typical environment for such applications includes high quality image capturing devices, high-end processing workstations, and recently, high bandwidth network connections for communications. The second observation is that the focus of these algorithms has primarily been the classification accuracy, where many methods have demonstrated excellent results in specific settings [12].

In [13] Yang et al. provided a systematic organization of face detection algorithms. The authors divided face detection algorithms into four classes. Knowledge-based methods consider heuristics, such as "a face is symmetric," "has two eyes," etc. Feature-based methods use color, texture, edges and shapes to identify faces in an image while trying to provide robustness to pose and lighting changes. Template-based methods identify areas that are highly correlated with either a predefined or a deformable face template. Appearance-based methods are similar to template matching, but rely more on algorithms from the machine learning community. The goal of the appearance-based methods is to identify the most discriminative features or functions of a face during the training stage and apply them during testing.

Among all four classes of algorithms, the appearance-based ones have become the most popular due to their excellent performance and potential for real-time implementation. Most of the appearance-based methods rely on sequentially searching through the target frame for patterns that closely resemble patterns that have been learned during the algorithm training. A wide range of machine learning algorithms have been used for learning the patterns associated with human faces and then trying to locate image subwindows that exhibit such characteristics. In [14], Sung and Poggio proposed a bootstrapping method for training a multilayer perceptron where misclassified samples are iteratively added to the training set and the model is retrained. In [15, 16], Rowley et al. used artificial neural networks based on spatial information of the face/non-face images. In [17] Osuna et al. were the first to use support vector machines for face detection. The method was later extended by using a cascade of detectors in [18]. In [19], the authors used convolutional neural networks and a bootstrapping technique for detecting faces. In [20], the authors used multiple Bayesian discriminating features, such as 1-D Haar wavelet coefficients and amplitude projections along with a statistic conditional probability density estimation method and achieved excellent results in diverse image datasets. One of the most successful methods for face detection is the Adaboost based scheme proposed by Viola and Jones [21], where a cascade of low complexity features offers significant computational savings without compromising the performance of the classifier.

Face detection in the compressed domain was first suggested by Wang and Chang [22], where the DCT coefficients were restored in their block form from JPEG or MPEG streams and then color based detection was applied. In [23], Fonseca and Nesvadha facilitated face detection in the DCT domain by utilizing information based on color, frequency and intensity. In [24], Luo and Eleutheriadis utilized color and texture in the DCT domain and used shorter feature vectors for color face classification.

3.1 Face Detection in Embedded Systems

Recently, there has been increasing interest on applying face detection algorithms in embedded systems. In [25], the authors investigated the implementation of face detection using Adaboost on a FPGA based Network on Chip (NoC) architecture.

NoC is a design paradigm borrowed from computer networks architectures that offers high detection rates by attaching duplicate modules. They reported that their system could achieve face detection in rates up to 40 frames per second and image resolution 320x240.

In [26], the authors experimented with Adaboost based FPGA implementation. The system consists of four elements: an image pyramid module responsible for scaling the subwindow, a FIFO buffer used for the calculation of the integral image, a piped register module for evaluating the subwindows and the classification module. They reported 86% detection accuracy on the MIT+CMU face database at 143 frames per second for images of size 640x480.

An FPGA implementation of the Viola-Jones face detection was presented in [27]. The authors proposed a parallel implementation of the algorithm that was able to provide real-time operation, supporting 30 frames per second in 640x480 pixel resolution images. In [28], Nair et al. proposed an embedded implementation of the Viola-Jones that operates in the compressed domain. The algorithm performs face detection on 216x288 pixel images at 2.5 frames per second. The authors also proposed an approximation to the integral image that was estimated directly from the DCT coefficients of the JPEG compressed images.

In [29], the authors presented a hardware implementation of the neural network based face detection proposed by Rowley et al. [15, 16]. The detection rate was an impressive 424 frames per second on 300x300 images. They reported an 8% drop in detection accuracy compared to the software implementation of the algorithm, but the results could be improved by operating on lower frame rates.

4 Face Detection under Power Constraints

Most of the face detection algorithms have been designed with respect to two primary objectives: maximize the classification accuracy and achieve real-time performance. Recent results indicate that these goals have been achieved under a variety of conditions. In general, computer vision applications involve high complexity operations including complex math, floating point operations and numerous memory transactions. Traditional computer vision systems assume high-end workstations, capable of performing millions of operations without any constraints in terms of power consumption.

Recent advancements in the fields of mobile systems and distributed smart camera networks present a new set of challenges for face detection. Limitations in processing capabilities, memory, communication bandwidth and available power are critical constraints that have to be taken into account in order to maximize the system lifetime. Hardware architectures for face detection, such as Field Programmable Gate Arrays (FPGAs), can offer high processing capabilities without requiring the power consumption or the cost of high-end workstations. In most of these systems, compression is a key step in the processing chain. It is applied in order to reduce the power necessary for transmission or memory for storage. Consider the following illustrative example from [3]. A NTSC QCIF resolution (176x120) requires 21 Kbytes. At 30 frames per second, the video stream requires 5 MBits/sec. The transmission rate for the IEEE 802.15.4, a network protocol that is widely used in

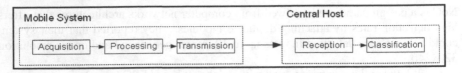

Fig. 2. Block diagram of the system

wireless video sensor networks, can offer up to 250 Kbits/sec. From this example it is apparent that image compression has to be applied, especially in situations where bandwidth is limited.

In this work, we consider the scenario where mobile resources are primarily dedicated to compression and transmission of the data to a central host. The detection/recognition algorithms are then applied at the central host after the image has been decompressed. Figure 2 shows the block diagram of this architecture. Image compression degrades the quality of the input image and introduces compression artifacts. In Section 6, we further explore how a face detection algorithm is affected by such degradations.

In our investigation, a set of assumptions were made. The first one is that we are only considering grayscale images. While color can be a valuable attribute in terms of face detection, it is more computationally expensive and requires the transmission of more data. In addition, the face detection methods we are investigating were developed for grayscale images and color can only offer minor gains in their detection accuracy. The second assumption is that the detection deals with frontal faces. Most of the algorithms were initially developed for frontal face detection and were later extended to pose invariant detection. In this sense, frontal face detection is the first step towards a more generic vision system that may include face recognition and tracking.

5 Machine Learning for Face Detection

5.1 Support Vector Machines Based Face Detection

Support vector machine (SVM) classifier is a machine learning algorithm that determines the hyperplane that linearly separates the training data with the maximum possible margin [30]. Formally, let the input training data be represented by a set of m labeled examples $x_i \in \mathbb{R}^d$, $i = \{1,2,\dots m\}$ and their associated labels $y_i = \pm 1$. The optimal hyperplane is obtained by maximizing a convex quadratic programming problem given by the following equation:

$$W(\alpha) = \max \sum_i a_i - \frac{1}{2} \sum_{i,j} a_i a_j y_i y_j \langle x_i, x_j \rangle. \tag{1}$$

subject to $\sum_i a_i y_j = 0$ and $0 \le a_i \le C$ for i=1...m, where a_i are the Lagrange multipliers of the optimization. The training samples associated with the non-zero Lagrange multiplies are called support vectors. A test vector $\tilde{x}$ is classified according to the decision function given by:

$$D(\tilde{x}) = w\varphi(\tilde{x}) + b = \sum_i a_i y_j \langle x_i, \tilde{x} \rangle + b. \tag{2}$$

In cases where the data cannot be linearly separated, the kernel trick is employed to transform the input data into a higher dimensional space where the transformed data is linearly separable. In this case the inner products in (1) and (2) are substituted by inner products in a higher dimensional space $\langle \Phi(x_i), \Phi(x_j) \rangle$. Unfortunately, evaluating all the inner products in the Φ space requires great computational effort. A solution to the problem is given by Mercer's theorem, where the inner products are given by a kernel function such that $K(x_i, x_j) = \langle \Phi(x_i), \Phi(x_j) \rangle$. The modified optimization function is then given by:

$$W(\alpha) = \max \sum_i a_i - \frac{1}{2} \sum_{i,j} a_i a_j y_i y_j K(x_i, x_j). \tag{3}$$

And the decision function is given by:

$$D(\tilde{x}) = w\varphi(\tilde{x}) + b = \sum_i a_i y_j K(x_i, \tilde{x}) + b. \tag{4}$$

There are various kernels that can be used in order to estimate high dimensional inner products. In this chapter, we use the second order polynomial kernel that has been successfully utilized in face detection and is given by $K(x_i, x_j) = (x_i \cdot x_j + 1)^2$.

SVMs were first used for face detection by Osuna et al. [17]. Later Heisele et al. [18], extended the method by introducing a hierarchy of SVM based classifiers, so that patterns that belong to the background are quickly rejected. In addition, a principal components based analysis was employed for selecting only the most discriminative features in the input space and a weighting scheme was applied to the feature space based on ranking the support vectors. In [31], Papageorgiou et al. used the wavelet representation as an input to a support vector machine based classifier for face and people detection in cluttered scenes. In [32], Shih et al. combined the input image, its 1-D Haar wavelet coefficients and its amplitude projection in order to derive a distribution-based measure for separating the class that was then combined with SVM and used for face detection.

5.2 Cascaded Adaboost Based Face Detection

An efficient method for detecting faces was proposed by Viola and Jones [21]. The core machine learning algorithm is based on the Adaboost framework. According to the framework, while a strong classifier might be difficult to obtain, it is much easier to obtain a sequence of weak classifiers that perform slightly better than random guessing. The combination of the weak classifiers (boosting) provides a strong classifier. Each weak classifier is tasked to learn one of three Haar-like features; a two-rectangle feature, a three-rectangle feature or a four-rectangle feature in a 24x24 pixel subwindow. Examples of such features are shown in Figure 3.

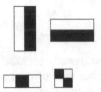

Fig. 3. Examples of rectangular features used in Viola-Jones face detection

In a given subwindow there are 160,000 possible features that have to be calculated. In order to manage the huge amount of features, the classifiers are cascaded. First, a simple classifier is trained. Then a second classifier is trained, with the inclusion that the inputs that were mistakenly classified by the first classifier are weighted more. The two classifiers are then combined and a third classifier is trained using the same weighting scheme.

In order to achieve real-time performance, the algorithm utilizes two characteristics. First, if the combination of the weak classifiers (one for each feature) is below a threshold for a test subwindow, the algorithm performs early rejection. Since most of the subwindows in an image are negative (non-face), the algorithm rejects as many subwindows as possible at the earliest stage. In addition, the computational complexity is significantly reduced by operating in a new image representation, called the *integral image*. At any given location (x,y) the integral image at that location is calculated as:

$$ii(x,y) = \sum_{x' \leq x, y' \leq y} i(x',y') \tag{5}$$

In the integral image, evaluation of the simple features amounts to a small number of additions/subtractions.

6   Face Detection in Images Degraded by Compression Artifacts

In this section we explore how the degradation of image quality due to compression artifacts affects the classification accuracy. The images were compressed using JPEG, JPEG2000 and SPIHT at various compression rates and the classification accuracy of two face detection algorithms, namely SVM and the Viola-Jones, were measured. The SVM classifier training was based on the sequential minimal optimization in Matlab [30]. The Viola-Jones testing was performed using the OpenCV edition of the algorithm for frontal face detection [33]. For JPEG compression, we used the built-in functionality provided by Matlab. For JPEG2000, we utilized the JasPer [34] encoder and the Matlab code for SPIHT was provided by M. Sakalli and W. A. Pearlman [35].

Training of both the SVM as well as the Viola-Jones classifiers was performed at the highest possible resolution, i.e. using uncompressed images. During testing, images were first compressed at various compression levels and then they were decompressed for face detection in the spatial domain. The classification accuracy for minimal compression levels corresponds to compression quality of 100 or compression rate of 1. The results in this case represent the best possible classification accuracy and serve

Fig. 4. Examples of faces in dataset

as an upper bound, since there are no compression artifacts to degrade classification performance. It is assumed that the classification error on uncompressed images is due to the limitations of the machine learning algorithm. Classification results obtained when higher compression rates are applied on the testing set can be used to determine the effects of compression on the classification error.

The dataset was provided by P. Carbonetto [36] and contains 4,196 images with human faces (positive examples) and 7,872 images that do not contain a face (negative examples). According to the author, the dataset is very similar to the one used during the training of the Viola-Jones classifier. The images are grayscale, 24x24 pixels with normalized variance. The 12,788 images in the dataset were divided into a testing set of about 8,485 images of faces and non-faces and a testing set of about 4,303 images of faces and non faces. Examples of the dataset images are shown in Figure 4.

Multiple trails were conducted for each point in order to achieve robustness to the individual training-testing pair. The same dataset was used in the training of the SVM classifier in order to reduce any variation in performance due to different training sets. Since the SVM was trained on the dataset, it is expected to achieve higher detection rates compared to Viola-Jones. We should note that the aim of this chapter is not to make a general comparison between the two algorithms, but to investigate and compare the robustness of each one when classifying images degraded by compression artifacts.

6.1 Classification Results for JPEG Compressed Images

We begin by presenting the results obtained using JPEG compression. The detection accuracy as a function of compression quality for SVM based classification is shown in Figure 5 (a). We observe that the SVM's detection accuracy is not compromised for quality settings above 30. There is a small drop in performance in very low quality settings, where the accuracy drops to 5% relatively to the uncompressed case. With SVMs, the classification is performed by evaluating the decision function shown in (4). Let us consider an input vector that would be correctly classified in the uncompressed case. The effect of the compression is a translation and/or scaling of the vector in the feature space induced by the kernel. The classification can still be performed without errors as long as the translated/scaled vector lies in the correct

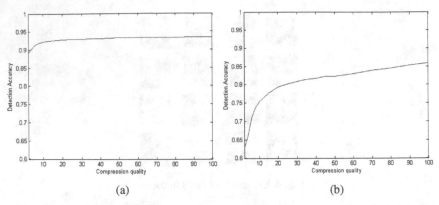

(a)   (b)

Fig. 5. Detection accuracy face detection in JPEG compressed images using SVM classification (a) and Viola-Jones classification (b)

hyperplane. Experimental results suggest that the large margin classification philosophy of the algorithm can provide robustness to testing data that have been degraded by compression.

The results of Viola-Jones classification on JPEG compressed images are shown in Figure 5 (b). We observe that the effects of compression are more pronounced compared to SVM. For reduced compression quality settings, there is a decrease in detection accuracy. More specifically, the decrease is almost "linear" for compression quality up to 20, where there is about 7% decrease in performance compared to the full quality.

The explanation of this significant decrease in detection accuracy can be found in the nature of the artifacts introduced by JPEG compression. More specifically, the quantization of the high frequency DCT coefficients, especially in low compression quality settings, introduces "ringing" artifacts that blur sharp edges [37]. The Viola-Jones algorithm relies on the detection of Haar-like features that are mostly represented by high frequency information, which is lost during quantization. As a result, the detection accuracy of the classifier is severely compromised by compression.

6.2 Classification Results for JPEG2000 Compressed Images

In this set of experiments, we investigate the effects of artifacts introduced by JPEG2000, on the detection accuracy of face detection using the SVM and the Viola-Jones classifier. The results for SVMs are presented in Figure 6 (a) and for Viola-Jones in Figure 6 (b). Regarding SVM classification, we observe that the detection accuracy is not compromised even in very low compression rates. The Viola-Jones algorithm is also robust to JPEG2000 compression, although there is a decrease in detection accuracy at very low compression rates (less than 0.2).

We observe that for both SVM and Viola-Jones the variation in classification error is negligible for rates above 0.2. In fact, similar to results obtained for face recognition, the detection accuracy is marginally increased for moderate compression

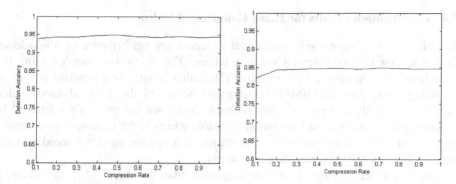

Fig. 6. Detection accuracy for face detection in JPEG2000 compressed image using SVM classification (a) and Viola-Jones classification (b)

rates. Compared to JPEG, JPEG2000 does not introduce significant "ringing" artifacts in the image because of its full-frame processing. It does introduce a blurring effect due to the attenuation of high frequencies. Fortunately, because of its supreme performance, the classification ability of the classifiers is not compromised.

6.3   Classification Results for SPIHT Compressed Images

In the last set of experiments, we investigate the effects of SPIHT compression on face detection. The results for SVM classification are shown in Figure 7 (a) and for the Viola-Jones in Figure 7 (b). We observe that similarly to JPEG2000 the performance of both the SVM and the Viola-Jones classifiers is not compromised by SPIHT compression. In fact, as we can see in Figure 10, the Viola-Jones performed better on images compressed at 0.5, than the uncompressed case. These results suggest that the wavelet based transform coding used in JPEG2000 and SPIHT does not introduce artifacts that can degrade the classification efficiency.

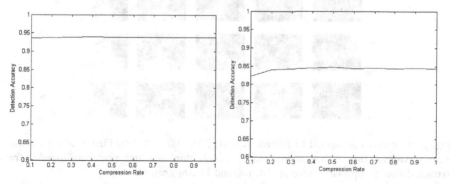

Fig. 7. Detection accuracy of face detection in SPIHT compressed images using (a) SVM based classification and (b) Viola-Jones based classification

6.4   Classification Results for H.264 Compressed Video

A series of experiments were performed to assess the performance of Viola-Jones face detection in a compressed video sequence. The video was encoded using the H.264/MPEG-4 Advanced Video Coding [39] video compression standard at various encoding rates. The H.264/MPEG-4 standard is one of the most advanced video encoding methods in terms of rate/distortion. Since we are primarily interested in mobile systems, we selected the baseline profile which is the minimum requirement profile and is less computationally demanding, as it ignores more advanced features such as B-slices, CABAC encoding and others.

The test video is part of the test sequence "david indoor" [40] and shows a person moving from a dark room to a bright room. This makes the sequence very challenging for face detection due to the large variation in illumination and large global and local motion. The frames are 320x240 pixels size. Figure 8 shows frames extracted from the sequence at various levels of compression. In all cases, the video was first compressed at the given encoding rate and then decompressed before extracting the frames. The most significant artifact introduced by the compression is the blocking effect that can be readily seen in the last row in Figure 8.

Figure 9 presents the results of the Viola-Jones face detector on H.264 encoded video. These results suggest that, as was the case with still images, there is a threshold on the encoding rate above which classification accuracy is not significantly affected. It should be noted that the maximum value of encoding rate, 884 Kbps, is the minimum possible compression using H.264/MPEG-4 AVC for this sequence. This observation explains the unnoticeable reduction in image quality between rows one and two in Figure 8.

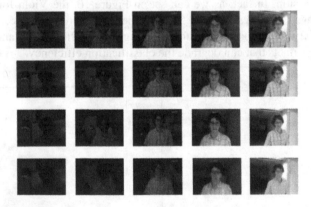

Fig. 8. Columns correspond to frames 1, 100, 200, 300 and 400. First row corresponds to frames extracted from uncompressed video. Second, third and forth rows correspond to frames extracted from compressed video at 884, 140 and 12 Kbps respectively.

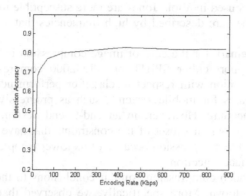

Fig. 9. Viola-Jones face detection performance on H.264 compressed video

7 Discussion

Recent advances in embedded systems provide new opportunities for computer vision algorithms. In this work, our attention was focused on face detection because, in addition to its own significance, it is considered to be an initial step to many high level computer vision algorithms, such as recognition and tracking. Migrating traditional face detection algorithms to resource constrained environments should be done with consideration to the limitations imposed by the environment including limited processing capabilities, memory, bandwidth and available power.

In this chapter, we explored the effects of image compression with respect to the detection accuracy of two face detection algorithms, using the SVM classifier and the Viola-Jones method. Three image compression techniques were examined, JPEG, JPEG2000, and SPIHT. Each compression technique is characterized by different qualities with regards to resource constrained environments. JPEG offers the lowest computational complexity at the expense of moderate compression capabilities. JPEG2000 provides excellent compression quality at the expense of increased complexity and SPIHT may be regarded as an intermediate method in terms of complexity and compression efficiency.

Experimental results suggest that JPEG greatly affects the detection accuracy of the Viola-Jones classification at all compression quality settings. SVM on the other hand maintains its detection accuracy for a wide range of compression settings. JPEG2000, a technique known for its good compression efficiency exhibited the minor degradation in classification performance for the Viola-Jones and a slight increase in detection accuracy for the SVM based. SPIHT also produced excellent results in terms of detection accuracy, with almost no effect on the accuracy of the SVM based and a marginal increase for the Viola-Jones. These results suggest that wavelet-based compression provides greater robustness to classification compared to discrete cosine transform based.

Experimental results indicate that the SVM based face detection algorithm is more robust to artifacts introduced by compression. This observation applies to all three compression methods under investigation. The robustness of SVM is attributed to the large margin philosophy of the classifier, imposed during training. On the other hand, the Haar-like features used in Viola-Jones are more susceptible to detection accuracy degradation, since they are described by high frequencies that are usually discarded during compression.

The concluding remark for the case of image compression is that if a system has the resources to support either SPIHT or JPEG2000 compression, it should be considered the best option with respect to classifier performance. In addition, they offer important features for mobile systems, such as progressive transmission and region-of-interest encoding. However, in an end-to-end design, image quality and detection accuracy are only a subset of the constraints that have to be met. Limited resources point to JPEG compression because of its lower complexity and acceptable accuracy with SVM face detection.

For the case of video compression, similar conclusions to the ones obtained for still images can be drawn. More specifically, we observed that face classification accuracy is not significantly compromised at low encoding rates even when degradation in image quality is severe. Experimental results suggest that even though the computational complexity of modern video encoders, such as H.264/MPEG-4 AVC is prohibiting for resource constrained systems, more lightweight encoders would be of value in resource constrained environments where computing power and bandwidth are primary considerations.

References

1. Abowd, G.D., Mynatt, E.D.: Charting Past, Present, and Future Research in Ubiquitous Computing. Transactions on Computer-Human Interaction, 29–58 (2000)
2. Rinner, B., Wolf, W.: An Introduction to Distributed Smart Cameras. Proceedings of the IEEE, 1565–1575 (2008)
3. Akyildiz, I.F., Melodia, T., Chowdury, K.R.: Wireless Multimedia Sensor Networks: A Survey. IEEE Wireless Communications, 32–39 (2007)
4. Wallace, G.K.: The JPEG still picture compression standard. IEEE Transactions in Consumer Electronics, xviii–xxxiv (1992)
5. Skordas, A.N., Christopoulos, C.A., Ebrahimi, T.: JPEG2000: The upcoming still image compression standard. Pattern Recongition Letters, 1337–1345 (2001)
6. Said, A., Pearlman, W.A.: A new, fast, and efficient image codec based on set partitioning in hierarchical trees. IEEE Transactions on Circuits and Systems for Video Technology, 243–250 (1996)
7. Santa-Cruz, D., Grosbois, R., Ebrahimi, T.: JPEG 2000 Performance Evaluation and Assessment. Signal Processing: Image Communication, 113–130 (2002)
8. Delac, K., Grgic, M., Crgic, S.: Effects of JPEG and JPEG2000 Compression on Face Recognition. Pattern Recognition and Image Analysis, 136–145 (2005)
9. Pekhteryev, G., Sahinoglu, Z., Orlik, P., Bhatti, G.: Image Transmission over IEEE 802.15.4 and ZigBee Networks. In: IEEE International Symposium on Circuits and Systems, pp. 3539–3542. IEEE Press, Los Alamitos (2005)
10. Wu, H., Abouzeid, A.A.: Power Aware Image Transmission in Energy Constrained Wireless Networks. In: 9th International Symposium on Computers and Communications, pp. 202–207. IEEE Press, Los Alamitos (2004)

11. Lee, D.U., Kim, H., Tu, S., Rahimi, M., Estrin, D., Villasenor, J.D.: Energy-Optimized Image Communication on Resource-Constrained Sensor Platforms. In: 6th International Conference on Information Processing in Sensor Networks, pp. 216–255. ACM, New York (2007)

12. Saha, S., Bhattacharyya, S.S.: Design Methodology for Embedded Computer Vision Systems. In: Kisacanin, B., Bhattacharya, S., Chai, S. (eds.) Embedded Computer Vision, pp. 27–47. Springer, London (2009)

13. Yang, M.H., Kriegman, D.J., Ahuja, N.: Detecting faces in images: A survey. IEEE Transactions on Pattern Recognition and Machine Intelligence, 34–58 (2002)

14. Sung, K.K., Poggio, T.: Example-Based Learning for View-Based Human Face Detection. IEEE Transactions on Pattern Recognition and Machine Intelligence, 39–51 (1998)

15. Rowley, H.A., Baluja, S., Kanade, T.: Neural Network-Based Face Detection. IEEE Transactions on Pattern Analysis and Machine Intelligence, 23–38 (1998)

16. Rowley, H.A., Baluja, S., Kanade, T.: Rotation Invariant Neural Network-Based Face Detection. In: IEEE International Conference on Computer Vision and Pattern Recognition, pp. 38–44. IEEE Press, Los Alamitos (1998)

17. Osuna, E., Freund, R., Girosit, F.: Training support vector machines: an application to face detection. In: IEEE International Conference on Pattern Recognition and Computer Vision, pp. 130–136. IEEE Press, Los Alamitos (1997)

18. Heisele, B., Serre, T., Prentice, S., Poggio, T.: Hierarchical classification and feature reduction for fast face detection with support vector machines. Pattern Recognition, 2007–2017 (2003)

19. Garcia, C., Delakis, M.: Convolutional Face Finder: A Neural Architecture for Fast and Robust Face Detection. IEEE Transactions on Pattern Analysis and Machine Intelligence, 1408–1426 (2004)

20. Liu, C.: A Bayesian Discriminating Features Method for Face Detection. IEEE Trans. Pattern Analysis and Machine Intelligence, 725–740 (2003)

21. Viola, P., Jones, M.J.: Robust Real-Time Face Detection. International Journal of Computer Vision, 137–154 (2004)

22. Wang, H., Chang, S.F.: A highly efficient system for face region detection in mpeg video. IEEE Transactions on Circuits and Systems for Video Technology, 615–628 (1997)

23. Fonseca, P., Nesvadha, J.: Face Detection in the Compressed Domain. In: IEEE International Conference on Image Processing, pp. 285–294. ACM, New York (2004)

24. Luo, H., Eleftheriadis, A.: Face Detection in the Compressed Domain. In: 8th ACM international conference on Multimedia, pp. 285–294. ACM, New York (2000)

25. Lai, H., Marculescu, R., Savvides, M., Chen, T.: Communication-Aware Face Detection Using NOC Architecture. In: Gasteratos, A., Vincze, M., Tsotsos, J.K. (eds.) ICVS 2008. LNCS, vol. 5008, pp. 181–189. Springer, Heidelberg (2008)

26. Lai, H.-C., Savvides, M., Chen, T.: Proposed FPGA Hardware Architecture for High Frame Rate (>>10 fps) Face Detection Using Feature Cascade Classifiers. In: 1st IEEE International Conference on Biometrics: Theory, Applications, and Systems, pp. 1–6. IEEE Press, Los Alamitos (2007)

27. Hiromoto, M., Nakahara, K., Sugano, H., Nakamura, Y., Miyamoto, R.: A Specialized Processor Suitable for AdaBoost-Based Detection with Haar-like Features. In: Embedded Computer Vision Workshop, pp. 1–8. IEEE Press, Los Alamitos (2007)

28. Nair, V., Laprise, P.O., Clark, J.J.: An FPGA-Based People Detection System. Journal on Applied Signal Processing, 1047–1061 (2005); EURASIP

29. Theocharides, T., Link, G., Narayanan, V., Irwin, M.J., Wolf, W.: Embedded Hardware Face Detection. In: Proceedings of the International Conference on VLSI Design, pp. 133–138. IEEE Press, Los Alamitos (2004)
30. Scholkopf, B., Smola, A.J.: Learning with Kernels. MIT Press, Cambridge (2002)
31. Papageorgiou, C.P., Oren, M., Poggio, T.: A general framework for object detection. In: 6th International Conference on Computer Vision, pp. 555–562. IEEE Press, Los Alamitos (1998)
32. Shih, P., Liu, C.: Face detection using discriminating feature analysis and Support Vector Machines. Pattern Recognition, 260–276 (2002)
33. Intel, Open Computer Vision library,
    http://sourceforge.net/projects/opencvlibrary/
34. JASPER Software Reference Manual, ISO/IEC/JTC1/SC29/WG1N2415
35. SPIHT in MATLAB Programming Language,
    http://www.cipr.rpi.edu/research/
36. Carbonetto, P.: Face Detection Dataset,
    http://www.cs.ubc.ca/~pcarbo/viola-traindata.tar.gz
37. Oztan, B., Malik, A., Fan, Z., Eschbach, R.: Removal of Artifacts from JPEG Compressed Document Images. In: Proceedings of SPIE, the International Society for Optical Engineering, pp. 1–9. SPIE (2007)
38. Misra, S., Reisslein, M., Xue, G.: A survey of multimedia streaming in wireless sensor networks. IEEE Communications Surveys and Tutorials, 18–39 (2008)
39. Sullivan, G.J., Wiegnad, T.: Video compression - from concepts to the H.264/AVC standard. Proceedings of the IEEE 93(1), 18–31 (2005)
40. Ross, D.: David Indoor, http://www.cs.toronto.edu/~dross/ivt/

Speech Recognition on Mobile Devices

Zheng-Hua Tan and Børge Lindberg

Multimedia Information and Signal Processing (MISP), Department of Electronic Systems,
Aalborg University, Aalborg, Denmark
{zt,bli}@es.aau.dk

Abstract. The enthusiasm of deploying automatic speech recognition (ASR) on mobile devices is driven both by remarkable advances in ASR technology and by the demand for efficient user interfaces on such devices as mobile phones and personal digital assistants (PDAs). This chapter presents an overview of ASR in the mobile context covering motivations, challenges, fundamental techniques and applications. Three ASR architectures are introduced: embedded speech recognition, distributed speech recognition and network speech recognition. Their pros and cons and implementation issues are discussed. Applications within command and control, text entry and search are presented with an emphasis on mobile text entry.

Keywords: Distributed speech recognition, embedded speech recognition, network speech recognition, mobile devices, text entry.

1 Introduction

ASR is a technology that converts a speech signal, captured by a microphone, to a sequence of words. After several decades of intensive research and with the help of increasing processing power of computers, the state-of-the-art speech recognition technology has already reached a level which enables the user to control computers using speech commands, to conduct dialogues with computers and to accurately dictate text (even when using large vocabulary continuous speech), mostly in controlled situations.

When placed in less controlled environments, such as noisy environments, speech recognition systems however encounter a dramatic performance drop. Noise robustness has long been a dominating research topic in the field of speech recognition. Sophisticated signal processing algorithms have been developed to effectively cope with noise and very robust systems for close-talking microphones are available nowadays.

These advances have heated up the enthusiasm of developing speech interfaces on mobile devices in which user interfaces are rather restricted by tiny keys and small displays [1]. Coupled with the restricted user interfaces, the increase in functionalities of the devices makes speech recognition on mobile devices an even more demanding challenge [2].

The progress of speech recognition technology in part relies on the increase in processing power of computers. Mobile devices, however, lack behind general-purpose

X. Jiang, M.Y. Ma, and C.W. Chen (Eds.): WMMP 2008, LNCS 5960, pp. 221–237, 2010.

computers for several years in particular in CPU-, floating point unit- and memory capacities [3]. A full-scale speech recognition system also requires a large amount of memory for storing acoustic models, pronunciation lexicon and language model which may go beyond the memory capacity of mobile devices. Furthermore, mobile device battery technology does not advance at the same astonishing speed as the microelectronics technology and therefore the amount of power available for the mobile devices is a significant limitation hindering the deployment of speech recognition on mobile devices. As a consequence, though speech recognition systems are optimized towards embedding on mobile devices [4], [5], the ASR performance is to some extent compromised. For example, it is too costly to run sophisticated noise robustness algorithms on a mobile device. Alternatively, given that most mobile devices have network connection, the speech recognition processing can be off-loaded to connected servers. This results in two new speech recognition architectures: network speech recognition [6], [7] and distributed speech recognition [8], [9].

Motivated by the maturing of speech recognition and the increase of computational power and network capabilities of mobile devices, start-up companies have recently emerged and entered the business of mobile speech recognition [10], [11]. This together with the large business enterprises within mobile technology and speech technology have pushed the technologies forward and made speech interfaces widely available on mobile phones, even on low-end devices. Speech applications have evolved from speaker-dependent name dialing that required user enrollment to speaker-independent systems that support dialing, command and control of phone functions, SMS dictation and even speech-enabled mobile search [12].

This chapter discusses the challenges and reviews the technologies behind the scene. It is organized as follows. Section 2 introduces automatic speech recognition and mobile devices, motivations to put them together and the challenges faced by speech recognition in the mobile context. Section 3 presents the fundamental and state-of-the-art techniques for mobile speech recognition. Section 4 presents applications and the chapter is summarized in Section 5.

2 Putting the Two Pieces Together

Mobile devices represent an enormously huge market. Every technology attempts to find applications in it, so does ASR. The difficulties in human-device interaction and mobile data entry provide an exciting opportunity for speech recognition. The driving forces for putting the two pieces together are as follows:

- Advances in ASR technology make many real-world applications feasible. Template based speech recognition was implemented in mobile phones more than a decade ago but unfortunately, due to the limitation of the technology, it was used on a daily basis merely by those who were experts in the technology. Nowadays, significant progress has been made in ASR. The use of statistical approaches eliminates the need of recording and training by the user and accuracy and robustness have been increased remarkably.
- Advances in mobile technology provide powerful embedded platforms and pervasive networking making the implementation of sophisticated ASR on mobile devices a reality either as an embedded or a distributed solution.

- The course of miniaturization limits the availability of user interfaces for mobile devices. Not everybody is comfortable with keypad and stylus - especially this varies across ages. Struggling with the low input bandwidth can turn a smartphone into a dumbphone used just for making and receiving calls for a large percent of the population; many advanced features and applications offered by the smartphone are thus untapped [13].
- Navigation in complex menu structures becomes inevitable but beyond manageable. Speech interface provides shortcuts that can make navigation much more efficient by avoiding going through the layered menu structure.
- Mobile devices are designed to be used while on the move, making interaction modes like keypad and stylus too awkward to use.
- An increasing number of countries enforce by law hands-free operations of the mobile phone in cars. A speech interface is well suited for hands-free and eye-free scenarios.

Michael Gold from SRI Consulting said, as much as TV transformed entertainment and PC transformed work, mobile technology is transforming the way that we will interrelate [14]. *The big questions here are whether speech technology will transform the way we interact with mobile devices and what need to be done to make it happen.*

To facilitate in-depth discussions, this section briefly presents the two pieces: automatic speech recognition and mobile devices.

2.1 Automatic Speech Recognition

In early days, dynamic time warping (DTW) was widely applied to isolated- and connected-word speech recognition especially in mobile devices. DTW is a pattern matching method with a nonlinear time-normalization effect where time differences between two speech signals are eliminated by warping the time axis of one so that the maximum coincidence is attained with the other, as the name indicates.

Although the template matching method is able to eliminate the problem related to varying speaking rate through dynamic programming alignment, it is incapable to characterize the variation among speech signals such as speaker variation and environmental noise due to the nature of being a non-parametric method. As a result, such a system is generally speaker dependent, fragile towards noise and with a small vocabulary. On the positive side, a DTW based speech recognition system is easy to implement and people with accents can train and use it. A recent work develops a Bayesian approach to the template based speech recognition [15].

ASR systems of today are primarily based on the principles of statistical pattern recognition, in particular the use of hidden Markov models (HMMs). HMM is a powerful statistical method of characterizing the observed data samples of a discrete-time series. The underlying assumptions for applying HMMs to ASR are that the speech signal can be well characterized as a parametric random process and that the parameters of the process can be estimated in a precise, well-defined manner [16].

The architecture of a typical ASR system, depicted in Fig. 1, shows a sequential structure of ASR including such components as speech signal capturing, front-end feature extraction and back-end recognition decoding. Feature vectors are first extracted from the captured speech signal and then delivered to the ASR decoder. The decoder searches for the most likely word sequence that matches the feature vectors on the basis of the acoustic model, the lexicon and the language model. The output word sequence is then forwarded to a specific application.

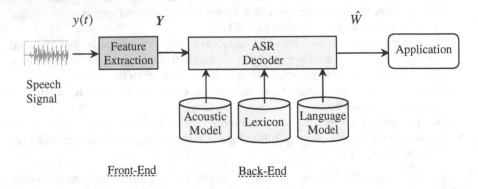

Fig. 1. Architecture of an automatic speech recognition system

Trained off-line with a large amount of data, an HMM-based ASR system can be speaker independent and is generally more robust against varying background noises as compared with a DTW system. On the other hand, the side effect of off-line training is that such a system is language and accent dependent. For each language or even each accent a system must be trained separately off-line and the training is rather costly in terms of time and data collection. An HMM-based ASR system is resource-demanding as well, which is not a problem for PCs of today, but a substantial problem for mobile devices.

2.2 Mobile Devices

A mobile device is designed to be carried and used while in motion or during pauses at unspecified locations. With significant technology advances in recent years, mobile devices of today are multi-functional devices capable of supporting a wide range of applications for both business and consumer use. For example, PDAs and smart phones provide the user not only with the functionality of making phone calls and sending SMS but also the access to the Internet for e-mail, instant messaging and Web browsing. Mobile devices are now functioning as an extension to the PC to enable people to work on the road or away from office and home. These rich features come along with cons as well. For example, manipulation becomes more difficult: As contact lists are becoming longer and longer, dialing a name from the address book will require on average 13 clicks [17].

As the size of mobile devices is small, there is little room for a keyboard and the screen size is always limited. A variety of interaction modes has been implemented for mobile devices, such as keypad, pen, haptics and speech. As being used while carrying, battery lifetime is always a big concern (around 3-5 hours in a mobile phone while talking). With more advanced features and applications, smartphones have a significantly shorter battery lifetime than dumbphones.

More importantly, mobile devices do not have as much computation and battery power and memory resources as personal computers have. Table 1 shows a comparison between an HP personal digital assistant iPAQ and an HP personal computer – both are off-the-shelf products from the Hewlett-Packard Company. The CPU of the PC is close to five times as fast as that of the PDA (clock-frequency wise). The RAM and the cache of the PC are more than two orders of magnitude larger. While floating-point units (FPUs) are available in some smartphones, the majority of them have to rely on floating-point emulation libraries. It is however rather slow and uneconomical to use the integer unit of a processor to perform floating-point operations and fixed-point implementation of algorithms is generally required to guarantee efficiency.

Table 1. Embedded platform vs. desktop PC

	CPU (clock freq.)	Arithmetic	RAM	Cache
HP iPAQ	624 MHz	Fixed-point	64 MB	16 KB
HP PC	3000 MHz, multi-core	Floating-point	8000 MB	6000 KB

Mobile platforms or operating systems (OS) that mobile devices are shipped with are dramatically diverse. According to Gartner Inc., in a decreasing order of worldwide market share in smartphones in 2008, the most widely adopted operating systems are: Symbian OS, RIM BlackBerry, Windows Mobile, iPhone OS, Linux and Palm OS [18]. The Android OS is now gaining market share as well. Due to the mass market, many mobile phones are shipped with proprietary operating systems as well. Java ME (J2ME) is the most used programming environment for proprietary operating systems.

2.3 Challenges in the Mobile Context

The challenges in deploying automatic speech recognition onto a mobile device are significant. The most notable challenge is the limited computational resources available on mobile devices as compared to desktop computers. To alleviate this problem, three types of mobile speech recognition have been developed: embedded speech recognition, network speech recognition and distributed speech recognition, which will be further discussed in the next section. For embedded speech recognition and distributed speech recognition, optimization for high efficiency and low memory footprint and fixed-point arithmetic are intensively researched.

In network speech recognition and distributed speech recognition, data transmission over networks takes place. Data compression and transmission impairments therefore need to be handled. When it comes to real-world deployment of ASR technology, algorithmic attention needs also to be given to the battery life and energy aware speech recognition becomes relevant to investigate as was done in [19]. Speech recognition by machine is not perfect and errors inevitably appear in the recognition results. However, error correction is a nontrivial problem, especially for mobile devices with a small screen. This highlights the importance of user interface design for both voice and GUI and of error correction through multiple modalities, which was shown to be more efficient in [20].

Diverse operating systems and hardware configurations bring about engineering issues and raise the cost for deployment. An example is speech data collection for mobile devices, as most speech databases available are collected using either telephones or high-quality headphones whose characteristics are considerably different from that of integrated microphones on mobile devices. Also, microphones on mobile devices are used in a different way and close-talking microphone setups are not convenient for the user. There is a significant lack of databases collected using mobile devices and the mismatch in training and test (application) acoustic data is known to degrade speech recognition performance. To compensate for these effects, Zhou et al. collected a large amount of speech data using PDA devices and general acoustic models are then adapted to the collected data to achieve more robust models [21]. Experiments on English dictation showed that using adapted models significantly improves recognition performance – word error rate decreases from 23.77% to 17.37%.

Mobile devices are often used in varying noisy environment and while on the move. Robustness against noise and degradation in speech quality is difficult to achieve under such conditions and in addition must be attained with limited resources.

A speech interface competes with existing and well-accepted user interfaces such as keypads and buttons. Relying on these by combining with a speech interface was found to be a good solution in [22].

3 Architectures and Techniques

Due to the challenges presented in the previous section, speech recognition on mobile devices is deployed with different architectures and for each architecture, the hindrances and the techniques to deal with them differ.

3.1 Three Architectures

To take advantage of the resources available from devices and networks, three approaches have been introduced: network speech recognition (NSR), distributed speech recognition (DSR) and embedded speech recognition (ESR) (see [23] for in-depth coverage). While ESR embeds the whole speech recognition system into the device, NSR and DSR make use of the connectivity available in the device and submit the entire speech recognition processing or part of it to a server.

3.1.1 Network Speech Recognition
NSR off-loads all recognition related tasks to a network-based server. Speech signals are in most cases encoded by a mobile or Voice-over-IP speech coder and transmitted to the server where feature extraction and recognition decoding are conducted as shown in Fig. 2. NSR in its basic form is a concatenation of two systems: a speech encoding-decoding system and a speech recognition system. This enables a plug and play of ASR systems at the server side while no changes are required for the existing devices and networks. The NSR approach has the advantage that numerous commercial applications are developed on the basis of speech coding. It further shares all the advantages of server based solutions in terms of system maintenance and update and device requirements.

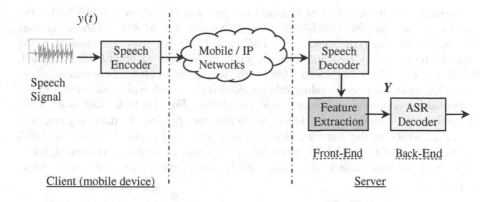

Fig. 2. Architecture of a network speech recognition system

The downside of NSR is network dependency. The user expects an application to run smoothly all the time and if network connection breaks, the resulted breakdown of NSR will bring a bad experience to the user. Further, speech coding and transmission may degrade the recognition performance due to such factors as data compression, transmission errors, training-test mismatch, etc. [24]. Among these factors, the effect of information loss over transmission channels has shown to be the most significant.

The degree to which a codec influences on ASR performance varies from codec to codec. For a tourist information task with a vocabulary of 5,000 words, the word error rate (WER) for the baseline system (i.e. without applying any codec) is 7.7% [49], while WERs for G.711 64 kbps and G.723.1 5.3 kbps are 8.1% and 8.8%, respectively. For the same recognition task, WERs for MPEG Layer3 audio codecs 64 kbps, 32 kbps, 16 kbps and 8 kbps are 7.8%, 7.9%, 14.6% and 66.2%, respectively. The results show that higher bit rate gives better recognition performance and that low-bit-rate audio codecs degrade ASR performance very significantly while speech codecs have rather limited influence.

Dedicated algorithms can be developed to eliminate the coding effect on ASR performance [6]. For example, the WER for a connected digits recognition task is 3.8% for wireline transmission and it is 5.3% for the IS-641 speech coder. To mitigate the effect of speech coding distortions, a bitstream-based framework has been proposed in which the direct transformation of speech coding parameters to speech recognition parameters is performed. The deployment of the bitstream-based framework decreases the WER from 5.3% to 3.8%. Algorithms of this kind are specific for each speech codec.

Although different databases have been employed for performance evaluation in the literature, the Aurora 2 database [50] is commonly used in this field and many results cited in the chapter are based on this database as well. The database is derived from the TI digits database and contains two training sets and three test sets. The three test sets (Set A, B and C) are contaminated by eight different types of noise with signal-to-noise ratio (SNR) values from 20 dB to -5 dB. Set C further includes convolutional noise. Clean speech is also included in the three test sets.

Experiments on the effect of transmission errors in [51] show that WERs for the Aurora 2 database for GSM-EFR (Global System for Mobile Communications Enhanced Full Rate) are 2.5%, 3.0%, 4.4%, and 12.9% for error-free channel, GSM Error Pattern 1 (EP1), EP2 and EP3, respectively. When a baseline DSR is applied, WERs are 2.0%, 2.0%, 2.1% and 9.0% for the four different channel conditions.

In [52] an uncertainty decoding rule for ASR is presented, which accounts for both corrupted observations and inter-frame correlation. For the task NSR over VoIP channels, a considerable improvement is obtained by applying the decoding rule. For example, WERs on the Aurora 2 database with simulated packet losses for G.729A and G.723.1 are 4.0% and 7.3% when only packet loss concealment is applied, while after applying uncertainty decoding, WERs are reduced to 2.9% and 3.6%, respectively.

3.1.2 Distributed Speech Recognition

As speech codec is generally developed on the basis of optimization for human perception, speech coding and decoding will degrade speech recognition performance. Instead of coded speech, speech features estimated for ASR can be directly quantized and transmitted through networks, which results in a new remote speech recognition architecture – distributed speech recognition [8], [9] as shown in Fig. 3. In the server the features are decoded and used for recognition. With recent advances in feature compression, channel coding and error concealment, the DSR approach both achieves a low bit rate and avoids the distortion introduced by speech coding [25], [26], [27], [28]. To provide the possibility for human listening, effort has also been put into the reconstruction of speech from ASR features with or without supplementary speech features such as pitch information and the results are quite encouraging [29]. The key barrier for deploying DSR is that it lacks foundation in the existing devices and networks that NSR has. Stronger motivation and more effort will be needed to make DSR grow in visibility and importance.

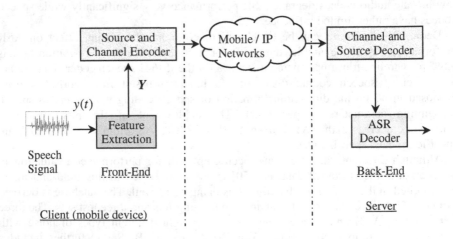

Fig. 3. Architecture of a distributed speech recognition system

A series of DSR standards have been published by ETSI. ETSI Standard ES 201 108 [30] is the basic front-end which defines the front-end feature extraction algorithm and the compression algorithm. ETSI Standard ES 202 050 [31] is the advanced front-end which includes a noise-robustness algorithm on top of the basic front-end. ETSI Standard ES 202 211 [32] is the extended front-end which includes a pitch extraction algorithm and a back-end speech reconstruction algorithm to enable speech reconstruction at the server side and tone-language speech recognition. Finally, ETSI Standard ES 202 212 [33] is the extended advanced front-end feature extraction algorithm that combines all the previous standards. A fixed point version of the extended advanced front-end is also available [34].

Research in DSR has been focused on source coding, channel coding and error concealment. A very promising source coding technique is a histogram-based quantization (HQ) [53]. It is motivated by the factor that acoustic noise may move feature vectors to a different quantization cell in a fixed VQ codebook and thus introduce extra distortion. The histogram-based quantization dynamically defines the partition cells based on the histogram of a segment of the most recent past values of the parameter to be quantized. The dynamic quantization method is based on signal local statistic, not on any distance measure, nor related to any pre-trained codebook. The method has shown to be efficient in solving a number of problems associated with DSR including environmental noise, coding and transmission errors. For Aurora 2 database (SetA, B, and C), the WERs are 38.9%, 43.5%, 40.1%, 22.8% and 18.7% for MFCCs (Mel-frequency cepstral coefficients) without quantization, SVQ (split vector quantization, applied in ETSI standards) 4.4 kbps, 2D DCT (two-dimensional discrete cosine transform) 1.45 kbps, HVQ (histogram-based vector quantization) 1.9 kbps, and HQ 3.9 kbps. In terms of noise robustness HQ has also shown to be better than a number of commonly used techniques such as MVA (mean and variance normalization and auto-regression moving-average filtering), PCA (principal component analysis) and HEQ (histogram equalization).

Another interesting approach is a front-end that uses *a posteriori* SNR weighted energy based variable frame rate analysis [54]. The frame selection is based on the *a posteriori* SNR weighted energy distance of two consecutive frames. It has been shown to be beneficial for both source coding and noise robustness. At 1.5 kbps with SVQ, the WERs are 1.2% and 32.8% for clean and noisy speech while WERs without comparison are 1.0% and 38.7%, respectively. The variable frame rate front-end has the potential to be combined with other methods such as the histogram-based quantization.

To achieve robustness against transmission errors, two classes of techniques have been applied: client-based techniques such as forward error correction, multiple description coding and interleaving, and server-based error concealment techniques such as insertion, interpolation, statistical-based techniques and ASR-decoder based techniques. These techniques and their pros and cons are extensively discussed in [9].

The WER performance on Aurora 2 task for a number of techniques is presented in Table 2. The ETSI standards (ETSI) apply a Golay code to protect the most important information e.g. the header information, use CRC (cyclic redundancy check) as the major error detection scheme and apply a frame repetition scheme as error concealment. ETSI standards are considered as the baseline. Weighted Viterbi decoding (WVD) is an ASR-server based technique in which exponential weighting factors based on the reliability of speech observations are introduced into the calculation of the likelihood. A subvector-level error concealment technique (SVQ-EC) is presented in [28] where subvectors in an SVQ are considered as an alternative basis for error concealment rather than the full vector and it is further combined with feature-based WVD. A novel uncertainty decoding (UD) rule is presented in [52] to take advantage of the strong correlation among successive feature vectors while considering the uncertainty of observation vectors in the HMM decoding process; the result for the inter-frame correlation based UD is cited from [52] whose baseline result differs from here. Interleaving24 represents an interleaving scheme in which a sequence of 24 vectors is grouped and interleaving is implemented simply by reading odd-numbered vectors first and even-numbered vectors second from the blocks and the approach introduces no delay when there are no transmission errors [28]. H-FBMMSE represents forward-backward MMSE with hard decision and the result is cited from [55]. In applying MDC, two descriptions are generated: the odd-numbered and the even-numbered, and the two descriptions are transmitted over two uncorrelated channels which both are simulated by EP3. The final one is WER without any transmission errors.

Table 2. WER (%) performance on Aurora 2 database corrupted by GSM EP3

Method	ETSI	WVD	SVQ-EC & WVD	Inter-Frame Correlation UD	Inter-leaving24	H-FB MMSE	MDC	Error-Free
WER	6.70	4.78	2.01	1.98	1.74	1.34	1.04	0.95

3.1.3 Embedded Speech Recognition

To be independent of network connectivity and avoid the distortion induced by compression and transmission detriments, ESR performs all ASR processing in the mobile device. Such a fully embedded ASR, however, consumes a large amount of computational resources and battery which all are scarce in mobile devices. This may not be tolerable as in most cases ASR is merely an integrated part of the user interfaces. Efficient implementation of algorithms is vital and can be realized through fixed-point arithmetic and algorithm optimization for reduced computation and small footprint. Speech recognition speed and resource consumption are now considered new metrics for measuring and comparing speech recognition systems [35]. In recent years, impressive ESR solutions have been developed [4], [5] and [10].

In addition, when the ASR involves large databases residing in networks, e.g. for compiling application specific grammars, bandwidth requirement and security concern turn out to be nontrivial. Update of the ASR engine is also inconvenient due to the widespread, numerous devices.

Calculating the state likelihood is one of the most time-consuming tasks in continuous-density HMM (CDHMM) based ASR systems and the calculation cost is dependent on the number of HMM Gaussians. One of the focused ESR research topics has therefore been on reducing the number of Gaussians while maintaining recognition performance. Multi-level Gaussian selection techniques are presented in [35] to reduce the cost of state likelihood computation. For a Broadcast News LVCSR task with a vocabulary of 118 000 words, the likelihood computation cost is reduced to 17% while the WER decreases by 3% in absolute value by using the selection methods together with efficient codebooks.

Memory efficiency is another important factor for ESR. Among others, memory efficient acoustic modeling is particularly interesting. In [56] around a 13-fold reduction in memory requirements was achieved for the Gaussian pool by converting the typical CDHMM-based acoustic models into subspace distribution clustering HMM (SDCHMM) based ones. For a 10K-word isolated speech recognition task, the baseline performance with a CDHMM-based recognizer is 4.1% WER, 36 x RT (real-time) and 6.3 MB Gaussian pool footprint, while the performance of a SDCHMM-based recognizer with fixed-point conversion and state-based K-nearest neighbor approximation for Gaussian selection is 4.8% WER, 2.5 x RT and 0.6 MB Gaussian pool footprint.

3.1.4 Comparison

Given the nature of mobile devices, the three ASR architectures for mobile devices will co-exist. There are pros and cons for each of the architectures and a comparison is provided in Table 3. Application scenarios are listed in the table as well whereas the boundary between them is indistinct, and the choice of which type to deploy will highly depend on the application and resources available.

Table 3. Comparison of the three architectures

	NSR	DSR	ESR
Network dependence	Yes	Yes	No
Transmission impairment	Yes	Yes	No
Coding effect	Yes	No	No
Computation complexity on device	Very low	Low	High
Memory footprint on device	Very low	Low	High
Battery consumption	Very low	Low	High
Portability	High	Medium	Low
Recognition performance	Compromised	Maintained	Compromised
Application scenarios	Telephone and VoIP applications, SMS dictation, voice search	Network based applications, SMS dictation, voice search	Command and control, SMS dictation, voice search

3.2 Systems and Implementation

This subsection presents a few systems and discusses their implementation issues.

It is very time consuming and requires high-level expertise to develop a fully embedded speech recognition system. As a result, most existing systems are proprietary. The CMU Sphinx-II is a well-known open source larger vocabulary continuous speech recognition (LVCSR) system which has been optimized and ported to mobile devices. The resulted system is called POCKETSPHINX, a free, real-time

LVCSR system for mobile devices [36]. The system runs on average 0.87 x RT on a 206 MHz mobile device.

Commercial systems have demonstrated satisfactory performance. An interesting work is the Vlingo system that uses hierarchical language modeling (HLM) and adaptation technology [11].

There are a number of systems based on the ETSI standards. As ETSI does not specify the protocol between the client and the server, developers have to propose and implement their own DSR protocol. DoCoMo developed a protocol running over HTTP [37]. Their DSR framework is used for applications such as e-mail dictation, route navigation and speech interpretation.

Extensive studies on DSR implementation issues related to J2ME, Symbian and Windows Mobile have been conducted in [38] and [39].

Based on the ETSI XAFE [34] and the SPHINX IV speech recognizer [40], a configurable DSR system is implemented in [38]. The system architecture is shown in Fig. 4.

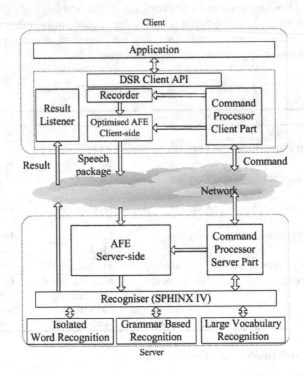

Fig. 4. System architecture of a real-time DSR system (From [38])

The DSR system supports simultaneous access from a number of clients each with their own requirements to the recognition task. The recognizer allows multiple recognition modes including isolated word recognition, grammar based recognition, and large vocabulary continuous speech recognition. The client part of the system is realized on a H5550 IPAQ with a 400MHz Intel® XScale CPU and 128 MB memory. Evaluation shows that conversion from floating-point AFE to fixed-point AFE reduces the computation time by a factor of 5 and most of the computation comes from the noise reduction algorithm deployed in the front-end and the feature calculation (MFCC) itself is computation light. With regard to memory consumption in the client, the size of the client DLL library file is only around 74 Kbytes, and the maximal memory consumption at run-time is below 29 Kbytes.

In [38] time efficiency comparisons were also conducted for the major components in the AFE: Noise suppression, waveform processing, computing MFCCs, post-processing and VAD processing. It was found that the majority of computation comes from noise suppression. On the PDA, the floating-point realization of the advanced front-end is 3.98 x RT while the fixed-point realization of the AFE achieves 0.82 x RT. The result indicates that the noise suppression of the AFE is computationally highly costly. The real time factor is further reduced to 0.69 x RT after using FFT optimization, which is in line with the finding of [57] where extensive experiments on the computational complexity of the AFE show that FFT computations account for a 55% of the total cost.

4 Applications

Applications for speech recognition can be classified into three major categories: command and control, search and text entry.

In terms of command and control the speech interface provides efficient shortcuts for bypassing the hierarchical menu structure so that making phone calls and launching applications become easier.

Speech enabled mobile search is a recent topic attracting much interest [58]. On mobile devices, speech recognition can be used to search in contact lists, applications, songs and the Internet. The application-oriented Voice Search Conference held annually is motivated by fast-growing services for mobile phones and the maturing of speech technology [41].

Mobile text entry represents a huge business sector: Over 2 trillion SMS messages were sent in 2008 world wide [18]. In addition, mobile devices are used for sending E-mails and instant messaging and for working on documents. Text entry is the focus of this section.

4.1 Mobile Text Entry

There are a variety of text entry methods. In a study conducted for entering text on a mobile phone in [42], speed data was obtained for two text input methods: T9 and multi-tap. When using multi-tap, the average words per minute (WPM) for both novice users and expert users are 8. When using T9, the WPM for novice users and for expert users are 9 and 20, respectively.

By using a prediction model, the work in [46] generates upper bound predictions for multi-tap and T9. The obtained results are cited in Table 4 and the numbers are much higher than that in [42] as the prediction model is used on the assumption of unambiguous words, expert users and no out-of-vocabulary words.

Human interaction speeds (potential text entry bandwidth) for a number of methods are presented in Table 4.

Note that human handwriting speed is up to 31 WPM, while the speed of machine recognition of handwriting is much higher. This means that the speed associated with handwriting is limited by humans, not machines [45].

Table 4. Human interaction speed

Interaction method	Word per minute
Multi-tap (timeout kill) [46]	25 (thumb), 27 (index finger)
T9 [46]	41 (thumb), 46 (index finger)
Handwriting [45], [47]	15-25 (general), 31 (memorized), 22 (copying)
Keyboard touch-typing [43]	150 (professional), 50 (average)
Speaking [43]	280 (conversation), 250 (reading),
Dictation [48]	107

4.2 Speech Enabled Text Entry

A study conducted in [48] shows that the average text entry rate was 107 uncorrected WPM when a large vocabulary continuous speech recognition system was used for dictation. Nevertheless, it took significantly longer time to perform error correction, which is of course highly dependent on the ASR accuracy.

A continuous speech recognition system for mobile touch-screen devices called Parakkeet is presented in [44]. They reported an average text entry rate of 18 WPM while seated indoors and 13 WPM while walking outdoors. Commercial systems, such as Nuance mobile text entry, are expected to significantly exceed this performance as the Nuance system outperformed the best human SMS typist.

As human speech has a communication bandwidth of up to 280 WPM [43], in a long-term perspective, ASR enabled text entry is a very promising alternative. A good short-term solution is to combine speech and keypad for efficient multimodal text input [22].

Although speech interfaces are natural and attractive due to advances in speech technology and wide use of mobile devices, many users are not using the available speech-enabled features. Cohen sums up three reasons for this: marketing, applications and interfaces and highlights that applications should be intuitive, work out-of-the-box and deliver value [12].

5 Summary

This chapter presented the challenges which the mobile speech recognition is facing and some techniques that deal with these challenges. Three architectures were discussed: network speech recognition, distributed speech recognition and embedded speech recognition. These architectures are expected to co-exist due to their different characteristics and application scenarios. A set of applications are reviewed as well.

References

1. Tan, Z.-H., Lindberg, B. (eds.): Automatic Speech Recognition on Mobile Devices and Over Communication Networks. Springer, London (2008)
2. Bailey, A.: Challenges and Opportunities for Intearction on Mobile Devices. In: Proc. COLING 2004 Robust and adaptive information processing for mobile speech interfaces, Geneva, Switzerland, August 2004, pp. 9–14 (2004)
3. Tan, Z.-H., Novak, M.: Speech Recognition on Mobile Devices: Distributed and Embedded Solutions. In: Tutorial at Interspeech 2008, Brisbane, Australia (September 2008)
4. Varga, I., Aalburg, S., Andrassy, B., Astrov, S., Bauer, J.G., Beaugeant, C., Geissler, C., Hoge, H.: ASR in Mobile Phones - an Industrial Approach. IEEE Transactions on Speech and Audio Processing 10(8), 562–569 (2002)
5. Novak, M.: Towards Large Vocabulary ASR on Embedded Platforms. In: Proc. ICSLP, Jeju Island, Korea (2004)
6. Kim, H.K., Cox, R.V.: A Bitstream-Based Front-End for Wireless Speech Recognition on IS-136 Communications System. IEEE Trans. Speech and Audio Processing 9(5), 558–568 (2001)
7. Peláez-Moreno, C., Gallardo-Antolín, A., Díaz-de-María, F.: Recognizing Voice over IP Networks: a Robust Front-End for Speech Recognition on the world wide web. IEEE Transactions on Multimedia 3(2), 209–218 (2001)
8. Pearce, D.: Robustness to Transmission Channel – the DSR Approach. In: Proc. COST278 & ISCA Research Workshop on Robustness Issues in Conversational Interaction, Norwich, UK (2004)
9. Tan, Z.-H., Dalsgaard, P., Lindberg, B.: Automatic Speech Recognition over Error-Prone Wireless Networks. Speech Communication 47(1-2), 220–242 (2005)
10. Cohen, J.: Is Embedded Speech Recognition Disruptive Technology? Information Quarterly 3(5), 14–17 (2004)
11. http://www.vlingo.com/ (accesed July 4, 2009)
12. Cohen, J.: Embedded Speech Recognition Applications in Mobile Phones: Status, Trends, and Challenges. In: Proceedings of ICASSP 2008, Las Vegas, USA (2008)
13. http://www.nuance.com/mobilesearch/ (accessed July 4, 2009)
14. http://www.thefreelibrary.com/ (accessed July 4, 2009)
15. Wachter, M.D., Matton, M., Demuynck, K., Wambacq, P., Cools, R., Compernolle, D.V.: Template-Based Continuous Speech Recognition. IEEE Transactions on Audio, Speech, and Language Processing 15(4), 1377–1390 (2007)
16. Rabiner, L.R.: A Tutorial on Hidden Markov Models and Selected Applications in Speech Recognition. Proceedings of the IEEE 77(2), 257–286 (1989)
17. http://www.nuance.com/devicecontrol/ (accessed July 4, 2009)
18. http://www.gartner.com/ (accessed July 4, 2009)
19. Delaney, B.: Reduced Energy Consumption and Improved Accuracy for Distributed Speech Recognition in Wireless Environments. Ph.D. Thesis, Georgia Institute of Technology (2004)
20. Suhm, B., Myers, B., Waibel, A.: Multi-Modal Error Correction for Speech User Interfaces. ACM Transactions on Computer Human Interaction 8(1), 60–98 (2001)
21. Zhou, B., Dechelotte, D., Gao, Y.: Two-Way Speech-to-Speech Translation on Handheld Devices. In: Proceedings of ICSLP 2004, Jeju Island, Korea (2004)
22. Hsu, B.-J., Mahajan, M., Acero, A.: Multimodal Text Entry on Mobile Devices. In: Automatic Speech Recognition and Understanding (ASRU), San Juan, Puerto Rico (2005)

23. Tan, Z.-H., Varga, I.: Networked, Distributed and Embedded Speech Recognition: An Overview. In: Tan, Z.-H., Lindberg, B. (eds.) Automatic Speech Recognition on Mobile Devices and Over Communication Networks, pp. 1–23. Springer, London (2008)
24. Peinado, A., Segura, J.C.: Speech Recognition Over Digital Channels. Wiley, Chichester (2006)
25. Bernard, A., Alwan, A.: Low-Bitrate Distributed Speech Recognition for Packet-Based and Wireless Communication. IEEE Trans. on Speech and Audio Processing 10(8), 570–579 (2002)
26. Ion, V., Haeb-Umbach, R.: Uncertainty Decoding for Distributed Speech Recognition over Error-Prone Networks. Speech Communication 48, 1435–1446 (2006)
27. James, A.B., Milner, B.P.: An Analysis of Interleavers for Robust Speech Recognition in Burst-Like Packet Loss. In: Proc. ICASSP, Montreal, Canada (2004)
28. Tan, Z.-H., Dalsgaard, P., Lindberg, B.: Exploiting Temporal Correlation of Speech for Error-Robust and Bandwidth-Flexible Distributed Speech Recognition. IEEE Transactions on Audio, Speech and Language Processing 15(4), 1391–1403 (2007)
29. Milner, B., Shao, X.: Prediction of Fundamental Frequency and Voicing from Mel-Frequency Cepstral Coefficients for Unconstrained Speech Reconstruction. IEEE Transactions on Audio, Speech and Language Processing 15(1), 24–33 (2007)
30. ETSI Standard ES 201 108; Distributed Speech Recognition; Front-end Feature Extraction Algorithm; Compression Algorithm, v1.1.2 (2000)
31. ETSI Standard ES 202 050: Distributed Speech Recognition; Advanced Front-End Feature Extraction Algorithm; Compression Algorithm (2002)
32. ETSI Standard ES 202 211: Distributed Speech Recognition; Extended Front-End Feature Extraction Algorithm; Compression Algorithm, Back-End Speech Reconstruction Algorithm (2003)
33. ETSI Standard ES 202 212: Distributed Speech Recognition; Extended Advanced Front-End Feature Extraction Algorithm; Compression Algorithm, Back-End Speech Reconstruction Algorithm (2003)
34. 3GPP TS 26.243: ANSI C Code for the Fixed-Point Distributed Speech Recognition Extended Advanced Front-End (2004)
35. Zouari, L., Chollet, G.: Efficient Codebooks for Fast and Accurate Low Resource ASR Systems. Speech Communication 51, 732–743 (2009)
36. Huggins-Daines, D., Kumar, M., Chan, A., Black, A.W., Ravishankar, M., Rudnicky, A.I.: POCKETSPHINX: A Free, Real-Time Continuous Speech Recognition System for Hand-Held Devices. In: Proc. ICASSP 2006, Toulouse, France (May 2006)
37. Etoh, M.: Cellular Phones as Information Hubs. In: Proc. Of ACM SIGIR Workshop on Mobile Information Retrieval, Singapore (2008)
38. Xu, H., Tan, Z.-H., Dalsgaard, P., Mattethat, R., Lindberg, B.: A Configurable Distributed Speech Recognition System. In: Abut, H., Hansen, J.H.L., Takeda, K. (eds.) Digital Signal Processing for In-Vehicle and Mobile Systems 2. Springer, New York (2006)
39. Zaykovskiy, D., Schmitt, A.: Deploying DSR Technology on Today's Mobile Phones: A Feasibility Study. In: André, E., Dybkjær, L., Minker, W., Neumann, H., Pieraccini, R., Weber, M. (eds.) PIT 2008. LNCS (LNAI), vol. 5078, pp. 145–155. Springer, Heidelberg (2008)
40. Lamere, P., Kwok, P., Walker, W., Gouvea, E., Singh, R., Raj, B., Wolf, P.P.: Design of the CMU Sphinx-4 Decoder. In: Proc. of Eurospeech (2003)
41. http://www.voicesearchconference.com/ (accessed July 4, 2009)

42. James, C.L., Reischel, K.M.: Text Input for Mobile Devices: Comparing Model Prediction to Actual Performance. In: Proceedings of the SIGCHI conference on Human factors in computing systems (2001)
43. Kolsch, M., Turk, M.: Keyboards without Keyboards: A Survey of Virtual Keyboards. University of California at Santa Barbara Technical Report (2002)
44. Vertanen, K., Kristensson, P.O.: Parakeet: a continuous speech recognition system for mobile touch-screen devices. In: ACM IUI 2009, Sanibel Island, Florida, USA (2009)
45. MacKenzie, I.S., Soukoreff, R.W.: Text Entry for Mobile Computing: Models and Methods, Theory and Practice. Human Computer Interaction 17(2), 147–198 (2002)
46. Silfverberg, M., MacKenzie, I.S., Korhonen, P.: Predicting Text Entry Speed on Mobile Phones. In: Proceedings of the CHI 2000 Conference on Human Factors in Computing Systems (2000)
47. Brown, C.M.: Human-computer interface design guidelines. Ablex Publishing, Norwood (1988)
48. Karat, C.M., Halverson, C., Horn, D., Karat, J.: Patterns of Entry and Correction in Large Vocabulary Continuous Speech Recognition Systems. In: CHI 1999 Conference Proceedings, pp. 568–575 (1999)
49. Besacier, L., Bergamini, C., Vaufreydaz, D., Castelli, E.: The Effect of Speech and Audio Compression on Speech Recognition Performance. In: IEEE Multimedia Signal Processing Workshop, Cannes, France (2001)
50. Hirsch, H.G., Pearce, D.: The Aurora experimental framework for the performance evaluation of speech recognition systems under noisy conditions. In: ISCA ITRW ASR 2000, Paris, France (2000)
51. Kiss, I.: A Comparison of Distributed and Network Speech Recognition for Mobile Communication Systems. In: Proc. ICSLP, Beijing, China (2000)
52. Ion, V., Haeb-Umbach, R.: A Novel Uncertainty Decoding Rule with Applications to Transaction Error Robust Speech Recognition. IEEE Transactions on Audio, Speech, and Language Processing 16(5), 1047–1060 (2008)
53. Wan, C.-Y., Lee, L.-S.: Histogram-Based Quantization for Robust and/or Distributed Speech Recognition. IEEE Transactions on Audio, Speech, and Language Processing 16(4), 859–873 (2008)
54. Tan, Z.-H., Lindberg, B.: A Posteriori SNR Weighted Energy Based Variable Frame Rate Analysis for Speech Recognition. In: Proc. Interspeech, Brisbane, Australia (2008)
55. Peinado, A., Sanchez, V., Perez-Cordoba, J., de la Torre, A.: HMM-based channel error mitigation and its application to distributed speech recognition. Speech Communication 41, 549–561 (2003)
56. Chung, H., Chung, I.: Memory Efficient and Fast Speech Recognition System for Low Resource Mobile Devices. IEEE Transactions on Consumer Electronics 52(3), 792–796 (2006)
57. Giammarini, M., Orcioni, S., Conti, M.: Computational Complexity Estimate of a DSR Front-End compliant to ETSI Standard ES 202 212. In: WISES 2009, Seventh Workshop on Intelligent Solutions in Embedded Systems, Ancona, Italy (2009)
58. Bacchiani, M., Beaufays, F., Schalkwyk, J., Schuster, M., Strope, B.: Deploying GOOG-411: Early Lessons in Data, Measurement, and Testing. In: Proceedings of ICASSP 2008, Las Vegas, USA (2008)

Developing Mobile Multimedia Applications on Symbian OS Devices

Steffen Wachenfeld, Markus Madeja, and Xiaoyi Jiang

Department of Computer Science, University of Münster, Germany
{wachi,m_made01,xjiang}@uni-muenster.de

Abstract. For the implementation of multimedia applications on mobile devices several platforms exist. Getting started with mobile programming can be difficult and tricky. This paper gives an introduction to the development of mobile multimedia applications on Symbian OS devices. In particular this paper presents the step-by-step development of an easy image processing application that makes use of the mobile device's camera. Here, a Nokia N95 phone is used, which is widely used in the community due to its features such as autofocus, GPS, wireless LAN, gravity sensors and more. This paper will provide background knowledge as well as step-by-step explanations of all implementation steps ranging from installation of an appropriate SDK to signing, installing, and running the developed application on a mobile device.

1 Introduction

This paper addresses the development of applications in Symbian C++ for the Symbian Operating System (Symbian OS). While there are quite many books on Symbian programming in general, e.g. by Harrison [7,8,9], the systematic coverage of programming multimedia applications is rather limited yet. Harrison and Shackman lately added one relatively short multimedia chapter in [9]. A good source for this topic is the book by Rome and Wilcox [11], which is fully devoted to multimedia on Symbian OS.

Many useful applications have been developed on Symbian OS mobile devices. The chapter by Wachenfeld et al. [18] contained in this book describes the reading of barcodes and its applications. Strumillo et al. [14] created a dedicated speech enabled menu, rather than a simple screen reader, for blind and visually impaired people. Along with ordinary phone functions (calls, SMSs) the programmed phone can be used as a speech recorder, a web browser, and a color recognizer in images captured by the phone's camera. Another application for the same group of impaired people is the localization of pedestrian lights described in Rothaus et al. [12]. The recent work of Ta et al. [10] describes an efficient tracking system using local feature descriptors. Bruns et al. [3] present an adaptive museum guidance system that uses mobile phones for on-device object recognition. In addition to these and many other works described in academic

X. Jiang, M.Y. Ma, and C.W. Chen (Eds.): WMMP 2008, LNCS 5960, pp. 238–263, 2010.

publications, a huge number of multimedia add-ons for Symbian OS devices are available in the commercial sector.

The goal of this paper is to help researchers and developers to implement multimedia applications on mobile Symbian OS devices. In particular this paper is a tutorial that shows step-by-step how to set up the hardware and the software development environment, how to implement a first application and how to make use of the camera and the autofocus feature. This paper has the form of a from-scratch tutorial, gives all necessary code examples, and provides several useful hints. Also, frequent errors will be addressed that developers may make when they are new to implementing mobile applications using Symbian OS. The main part of this paper is a step-by-step implementation of a mobile application that uses the device's camera, the autofocus feature, and which displays resulting images after a simple pixel-based image manipulation. Several Symbian OS powered devices exist and will be compatible with the examples given here. All examples have been tested on a Nokia N95 smart phone, which holds the greatest share of smart phones at the time of writing.

The remainder of this paper is structured as follows: Section 2 covers the proper setup of a development environment, covering both hardware and integrated development environment (IDE). Section 3 explains the necessary steps to implement a Symbian OS program, programming conventions, and the important mechanisms for exception handling and object construction. Section 4 is the main part and describes the step-by-step implementation of an application example by incrementally extending a 'Hello World' application. Section 5 gives a summary and concludes this paper.

2   Development Environment

While the number of programmable mobile devices is large, there is a common way to implement applications for them. In general, an Integrated Development Environment (IDE) is used, which can be be installed on a normal computer. Such IDEs comprise well-known standard IDEs such as Eclipse, Microsoft Visual Studio, Apple's Xcode or specific IDEs like Nokia Carbide.

IDEs generally have to be extended by a device-specific Software Development Kit (SDK). For most devices also an emulator exists, which allows to test the implemented application on a computer without using the mobile hardware itself. One specific problem of multimedia applications is that available emulators do not always emulate the device's multimedia capabilities, such as photo and video functionalities. This problem often leads to the requirement of installing and debugging multimedia applications on the mobile device itself.

Although this paper focuses on the development of mobile applications for Symbian OS devices in general, we will introduce the Nokia N95 device in particular. The N95 is currently the most commonly used Symbian OS device. We will distinguish it from other available mobile devices and explain how to set up the environment which is needed to develop applications for Symbian OS devices.

2.1 Hardware

Today's mobile devices have reached a computational power which allows them to perform quite complex tasks. Advanced mobile devices integrate mobile computers, handheld game consoles, media players, media recorders, communication devices, and personal navigation devices. Such devices are produced by various manufacturers and differ regarding their operating systems and the supported languages to program them.

For multimedia applications generally devices are preferred that have strong computational power, enough memory and a good integrated camera. For this tutorial we have selected the Nokia N95 smart phone. The Nokia N95 has an integrated 5 megapixel camera with an excellent Carl Zeiss lens and, most importantly, with autofocus. It is powered by a dual 332 MHz Texas Instruments OMAP 2420 processor which is powerful enough for most image processing tasks as well as simple real time processing of video stream data. The performance is sufficient to display the 640 × 480 pixel video stream with about 20 frames per second (fps) after applying a 3 × 3 image filter operation to it.

The camera supports three general modes for image and video capture:

- **Photo mode:** The photo mode allows for the acquisition of photos in a resolution of up to 2592 × 1944 pixels and for processing in memory in different formats. The photo mode allows for the use of autofocus. A disadvantage of this mode is that it requires some time to be activated.
- **Video mode:** The video mode acquires images in a physical resolution of 640 × 480 pixels at 30 fps. Their format is fixed to YUV420planar which has to be converted manually if other formats such as RGB are wanted. Another disadvantage of this mode is that it does not support the autofocus.
- **Viewfinder mode:** The viewfinder is very fast and acquires images in a physical resolution of 320 × 240 pixels. The images are in RGB format and can be displayed directly. The main advantage of this mode is that it combines video stream with the possibility to use autofocus.

All modes allow for the use of the digital zoom which works without reducing the acquisition speed. The high resolution and the autofocus capability make this mobile phone an excellent choice. Nevertheless, a big drawback is that the currently available libraries do not allow a direct control of the focus or the white balance.

Among the alternatives to Symbian OS devices are Blackberry smart phones, Apple's iPhones, Google's Android phone, and smart phones from other manufacturers such as Samsung or Palm. A reason to choose the N95 device as a platform for development is that in addition to the video-capable autofocus camera, this phone integrates GPS, Wireless LAN, and an acceleration sensor. This allows for a wide range of applications.

The currently available G3 Android Phone has similar features and the development of applications for it is described in [4]. The Android phone is especially interesting because all libraries are open source and can be extended by everybody. Also, it is one of the few phones whose photo and video functionalities can be tested and debugged on a PC using a web-cam and a software emulator. Currently only very few Android phones exists and the Android OS is still under development which means that some smaller features are still deactivated.

One of the most popular mobile devices is Apple's iPhone which exists as iPhone, iPhone 3G, and iPhone 3GS. Only the newest iPhone 3GS has video capability and autofocus. The pre-3GS iPhones have thus not been the first choice for imaging applications. But this might change, as the new iPhone 3GS which has autofocus and is video capable is sold in large quantities. A drawback is that the development of iPhone applications requires an Intel-based Macintosh computer running Mac OS X 10.5 or later and a fee-based registration to become a developer at Apple. Nevertheless, the iPhone 3GS has become a very attractive platform for mobile applications and can be expected to be very popular for imaging applications soon.

2.2 Setting Up an Integrated Development Environment

In the following we will describe how to attach the N95 phone to a computer, and how to install the tools necessary for on-device debugging, as for imaging applications it is currently necessary to install and debug them on the phone because there is no possibility to test the camera facilities (both photo and video mode).

Installation of the IDE. The N95 can be programmed using Java (J2ME), Python, and Symbian C++. Here, we will show how to use Symbian C++ which offers the highest speed and allows a very direct control of resources. In contrast, Java is slower and currently does not allow to access the video stream.

To implement applications in Symbian C++, a plug-in for Microsoft Visual Studio exists. But as the support for this plug-in was discontinued, we recommend using Carbide C++. Carbide C++ is an Eclipse based IDE and is available for free since version 2.0 [6]. In addition to the IDE the Symbian OS Software Development Kit (SDK) is necessary. The current version for the N95 phone which will be described here is S60 3rd Edition, Feature Pack 1. IDE and SDK can be installed on any PC running Windows XP or Windows Vista[1]. The SDK further needs a Java Runtime and (Active) Perl Version 5.6.1. Links to download this software will be shown within the Carbide IDE after installation. Perl needs to be version 5.6.1 exactly, as newer versions do not allow to compile help files for mobile applications properly.

Connecting the Symbian device to a PC. The N95 phone as well as most other Symbian devices can be connected to a PC using a USB cable. Also it is possible to establish a connection via Bluetooth. Before connecting the device it is recommended to go to the Nokia website and to install the newest driver.

---

[1] Carbide C++ does not officially support Windows 7 yet.

Fig. 1. Connection wizard - installation of TRK on the mobile device via Carbide

If the device is connected and properly recognized, the device's firmware should be upgraded to the newest available firmware which is available on the Nokia website. Firmware updates lead to an increased stability and remove several bugs. In the case of communication problems, make sure that the communication mode is set to 'PC Suite' on the mobile device.

To establish the connection to the IDE, Carbide offers a wizard which can be found under 'Help→On-Device Connections...'. Within this wizard the connection type and the port has to be selected. The correct port will show the name of the mobile device and is thus easy to spot.

Now, as a basic connection is established, a special software needs to be installed on the mobile device. The software is called TRK and stays in the background and communicates with the PC's IDE in order to allow on-device debugging. The TRK software can be installed on the mobile directly via the IDE. The appropriate 'Application TRK' has to be selected on the tab 'Install remote agents'. The correct version is that one whose folder name corresponds to the mobile device's operating system. The N95 phone runs S60 3rd edition, Feature Pack 1. Thus, the correct choice is the TRK with the folder name '3.1.0' (see Figure 1). The TRK installs immediately after clicking 'install' but can also be saved as a '.sis' file in order to copy and install it manually onto the device. Once installed, TRK can be launched on the mobile device in order to set or change the connection type and connection parameters. After these steps, the mobile device is properly connected and ready to communicate with the IDE.

3   Implementing Mobile Applications Using Symbian OS

This section covers general aspects of implementing mobile applications using Symbian OS. At first, steps will be explained which are necessary to build an application that can be installed and run on a Symbian OS device. These steps differ slightly from those of a normal software development process and they comprise extra steps such as 'signing'. Then, this section explains coding conventions that exist for Symbian C++, special mechanisms such as exception handling and safe object construction using two-phase constructors.

3.1   Process Overview

The process to run self-written code on the mobile device can be divided into five steps.

- **Compiling:** The source code has to be compiled to the device's machine code.
- **Linking:** In the linking step, all libraries or external functions that are used within the program are linked.
- **Packaging:** Packaging creates a single '.sis' file which can be installed on the mobile device. The '.sis' file contains all files that belong to the application, e.g. program files, resource files, image files or sound files.
- **Signing:** Since S60 3rd edition, the Nokia operating system only allows to install signed programs. A program's signature is requested as a security mechanism and includes details about the 'capabilities' the program will use. Some details about signing programs will be explained in the following subsection.
- **Installing:** The last step is to copy the signed '.sis' file and to install it on the device.

3.2   Compiling, Linking and Packaging

Within Carbide programs are written in Symbian C++. In order to run, such programs have to be compiled, linked and packaged. These steps are performed in an automated manner by Carbide. These steps can be configured using three important files, that belong to the project.

- **pkg-file:** The pkg-file determines how the '.sis' file will be packed together and it also determines the application's name and Secure ID (SID). The SID will be explained below.
- **mmp-file:** The mmp-file is the main configuration file of a Symbian OS project which will be used in the compiling and linking step. This file contains names, UIDs, target format ('.exe' or '.dll'), include paths which will be searched for header files, a list of all '.cpp' files that have to be compiled, the capabilities used by the application, and the linked '.lib' library files.

- **Bld.inf file:** The Bld.inf file is like a make-file that determines what will be executed in what order to create the application. Here, in a section called 'prj_mmpfiles' the mmp-file is specified that will be used.

To identify files on the device, the first 12 Bytes of each file contain information in the form of three UIDs. UIDs are explained in [1], but the following gives a brief overview.

- **UID1:** UID1 encodes the target format of the file. Valid formats are .exe executable, DLL file, direct file store (data, not executable), or permanent file store (data, not executable).
- **UID2:** UID2 has different meanings depending of the file it belongs to. It will be ignored for executable files that are non GUI-applications.
- **UID3:** UID3 has a unique value for each individual application program and is also used as the SID. The SID for a binary is not specified explicitly, instead it is taken to be the same as the third UID. If an application has a SID of zero, this means 'undefined', and has several consequences, including lack of privacy for data used by the application. All files which have the same UID3 are considered to belong to the same project. For different projects, different UID3s have to be selected.

  Note: A common problem occurs if implementing a new application is started by copying the code of another project without changing the UID3. Copying both the old and the new project to the mobile device when they still have the same UID3 will cause problems. In general all applications that try to use the same UID3 will interfere and cause problems such as not permitting to delete one application properly or to start the applications. Thus, always a new UID3 should be given to a new project.

3.3   Signing Applications

Symbian OS has a security mechanism that controls the access to sensible features and user data on the mobile device. Such features include the use of the camera, the file system or networking. If a program requires such functionalities, it has to declare these needs which are called 'capabilities' in its '.sis' file. The requested capabilities of a program will be listed and shown to the user at the time of installation together with the author who signed the '.sis' file and the time and validity of the signature. This way users have a way to understand where a program comes from and what capabilities it is going to use. If programs require unnecessary capabilities, e.g. if a downloaded game like Minesweeper requests access to the local address book or the SMS system, users should be skeptical and may decide not to install the eventually harmful code.

On Symbian devices which run S60 3rd edition or higher, only signed programs can be installed. For the development this means that a program has to be signed and re-signed every time it is recompiled. The different alternatives to sign applications are explained in [17]. Generally for unrestricted applications that can be copied to different devices the following two alternatives exist. For both alternatives a TC Publisher ID is required which developers can purchase for \$200 per year directly from the TC TrustCenter GmbH[2].

- **Certified sign:** To obtain a certified signed application the whole application has to be sent in to a company that will perform several tests and check whether the program works as described. To obtain price information a registration is required. For more information see [16].
- **Express sign:** The easier way is to obtain express signed applications, where the developer can perform the tests himself and just sends in the test results.

For the development a much more convenient way to sign exists, the so-called 'self-signing'. This means the application is signed by the developer himself without the need of having a TC Publisher ID. The advantage is that everyone can do this and that it facilitates the development. The disadvantage is that the created certificate is not trusted. This means upon installation a warning will be shown that the program is from an unknown and possibly harmful source. Further, using self-sign certificates restricts the use of capabilities. Applications can use only the following capabilities: 'LocalServices', 'NetworkServices', 'ReadUserData', 'UserEnvironment', and 'WriteUserData'.

The application and all files that belong to it are contained in an '.sis' file which is the result of the packaging step. Signing the application creates an '.sisx' file from the '.sis' file. This file can be copied and installed onto the mobile device. Carbide can automatically perform the copy and installation process. The '.sisx' file contains the SID which is the same as the UID3.

If the UID3/SID is not set or zero, files are not considered to belong to a specific application and are not protected any more. In particular this means they are accessible by other applications or may even be deleted.

3.4   Conventions

Like in other programming languages, several conventions exist for the programming in Symbian C++. Although some of them may appear special, it is highly recommended to follow these conventions. It will make life easier.

Carbide helps to follow these conventions by providing a 'CodeScanner'. The CodeScanner can be activated by right-clicking on a '.cpp' file within Carbide's project manager. The CodeScanner searches the source code, looks for errors and gives hints related to the programming conventions.

Class names. Class names should start with special capital letters. These letters indicate the location where these classes will be saved internally and the way they will be used.

- C-classes will be saved on the heap. This is the standard form of classes, e.g. `CMyWonderfulClass`.

---

[2] TC TrustCenter GmbH, Sonninstr. 24-28, 20097 Hamburg, Germany.
TC Publisher IDs: http://www.trustcenter.de/order/publisherid/de

- T-classes are classes which are used to save data, they are saved on the stack.
- R-classes are resource classes which hold a handle to a resource owned elsewhere, for example, in another local object, a server or the kernel. All the Symbian OS resource classes have their name preceded by an R so that they are easily recognized. A detailed introduction to R-classes is given in [15].
- M-classes are interfaces and as such they do not have an implementation.

Variable names. The names of variables begin with different letters as well that give some information about the variables. Class variables begin with a small i. Arguments of functions begin with an a. Constants begin with a capital K. After such a leading character the variable name is continued with a capital character, e.g. aWidth or KErrNone. Normal local variables start with a small character.

Methods. Method names always start with a capital letter. Set-methods will have a leading Set in their name. Get-methods that use a reference variable have a leading Get in their name, while get-methods that return values directly do not have a leading Get. Example:

```
class TRectangle {
public:
    TRectangle();
    TRectangle(TInt width, TInt height);
public:
    void SetWidth(TInt aWidth);
    void GetWidth(TInt& aWidth);
    TInt Width();

    void SetHeight(TInt aHeight);
    void GetHeight(TInt& aHeight);
    TInt Height();
private:
    TInt iWidth;
    TInt iHeight;
}
```

3.5 Exception Handling Using Leaves

Methods which may fail, e.g. because a capability or required memory is not available, are indicated by a capital L at the end of their name. The L stands for 'leave'. A leave can be catched in a so-called 'trap'. The leave/trap mechanism of Symbian OS is like the try/catch-mechanism for exceptions in Java. Example:

```
TInt error;
TRAP(error, fooFunctionL());
if (error!=KErrNone){
    // Treatment of exceptions thrown by "fooFunctionL()"
}
```

TRAP is a macro which requires a defined variable and a function. After successful execution of the function the variable will contain 0, in case a leave was caused due to an error the variable will hold an error code. As an alternative TRAPD can be used to catch leaves, which internally creates the variable that holds the error number. If leaves shall be caught but not treated, TRAP_IGNORE can be used. Further explanations of the leave/trap mechanism are given in [2,13].

The use of leaves can be illustrated well at the example of constructing objects. The following function is used to create two objects of some class.

```
void CTest::createObjectsL(){
   CSomeClass* obj1 = new (ELeave) CSomeClass;
   CSomeClass* obj2 = new (ELeave) CSomeClass;
   ...
}
```

The name createObjectsL indicates that this function may fail due to a leave. Using the key word (ELeave) causes such a leave in case the construction of the corresponding object fails. The memory reserved for that operation will be automatically freed. But only the memory of the operation is freed, not that of preceding successful operations. If obj1 was successfully created but the creation of obj2 fails, the allocated memory of obj2 is freed but that of obj1 is not of course. But as obj1 is of local scope, it cannot be used outside. This behavior may cause memory leaks.

To avoid this, the so-called 'CleanupStack' can be used which allows to add objects by using a push-method and removing them again by using a pop-method. The idea is to remember objects that are in construction on this stack. Upon complete creation of all wanted objects, the CleanupStack can simply be emptied and nothing happens. This is similar to a 'commit' in database systems. If something goes wrong and a leave is caused, all objects in the CleanupStack will be destroyed and their memory will be freed. In particular the function PopAndDestroy will be automatically called for all elements on the stack. This is conceptually like a 'rollback' in a database system. The following example shows how to put objects on the CleanupStack:

```
void CTest::createObjectsUsingCleanupStackL(){
   CSomeClass* obj1 = new (ELeave) CSomeClass;
   CleanupStack::PushL(obj1);
   CSomeClass* obj2 = new (ELeave) CSomeClass;
   CleanupStack::PushL(obj2);
   ...
}
```

Using this code will cause the memory of both objects to be freed by the TRAP macro if it catches a leave. In general it does not matter where a leave is caused or catched, the CleanupStack is a singleton and thus holds a global list of references to objects in memory that will be deleted. This means, that leaves caused by a subsequent instruction or operation will cause the memory to be freed as well. Thus, after successful object creations the references have to be removed from the

CleanupStack using the CleanupStack::Pop() method. The following example of two-phase constructors will also show the use of this method.

3.6   Two-Phase Constructors

The CleanupStack is also used to create so-called memory safe two-phase constructors. This is a way to deal with the fact that it is not possible to put objects on the stack between the memory allocation and the call of a constructor. To do this, the additional method ConstructL is used which holds code that would normally be located in the constructor. Moving the code to this method allows to save and cleanup the memory for the class object and for other objects created within the object. Example:

```
// Phase #1
CManager::CManager(){
    // No implementation required
}

// Phase #2
void CManager::ConstructL(){
... // Operations that construct the object
... // and that may cause a leave too
}

CManager* CManager::NewLC(){
    CManager* self = new (ELeave)CManager();
    CleanupStack::PushL(self);
    self->ConstructL();
    return self;
}

CManager* CManager::NewL(){
    CManager* self=CManager::NewLC();
    CleanupStack::Pop(); // self;
    return self;
}
```

This example shows the construction mechanism for the class CManager. Instead of creating a new instance using the new command, the static method NewL will be called:

```
CManager* manager = CManager::NewL();
```

The creation is done by calling the NewLC method. Within the NewLC method, the empty constructor is called to allocate the memory and to create the empty object (phase 1). Then the object is put on the CleanupStack. The real object creation can be performed in method ConstructL (phase 2). In case either the memory allocation or some operation within the ConstructL method fails, the CleanupStack is used to destroy the created objects and to free the memory. If NewLC succeeds, the CleanupStack is emptied and the NewL method returns a reference to the created object.

4   Step-by-Step Development Example

This section is the main part of this paper. Step-by-step the development of a mobile application will be described. The application will make use of the device's camera, be able to use the autofocus and apply a simple pixel-based image operation to the captured frames. Starting from a 'Hello World' application, the development will be explained in an incremental way. All important pieces of source code will be given and explained. The section ends with some additional tips and comments concerning further improvements.

4.1   Start: A 'Hello World' Application

The creation of a new application is very easy and can be done by extending existing code examples that can be generated using a wizard. The easiest way is to create a simple 'Hello World' application, that prints the words 'Hello World' on the mobile device's display. After opening Carbide C++ the menu 'File →New →Symbian OS C++ Project' starts the project wizard. From a list of different project types the type 'S60 Gui Application' should be chosen. All that needs to be done is to give a name to the project and to select an installed SDK. Leaving all other settings to the default and clicking 'Finish' creates the 'Hello World' example.

Without modifying any code, this example should be tested on the mobile device. First, it should be verified that the application is built and signed properly. This can be seen by looking at the projects properties (Right-click the project within Carbide's project explorer and select 'Properties'). Within the property window, select 'Carbide.c++' and then 'Build Configurations' (see Figure 2). By default there are five different build configurations which can be selected from the 'Configurations' drop-down menu. To create a debug version of an application that runs on the N95 and uses the video stream, 'Phone Debug (GCCE)' is the correct choice. All other configurations can be deselected using the 'Manage...' option. The 'SIS Builder' tab shows if the correct '.sis' file is going to be signed. If no entry exists, choose 'Add', select the '.pkg' file of the project from the 'sis' directory and sign it using 'Self sign sis file'. This is all that needs to be done, the properties can be closed using 'OK'.

To receive and run the application on the mobile device, TRK has to be started on the device and set to 'connect'. Then, within Carbide, the project can be right-clicked again and the menu 'Debug As →Debug Symbian OS Application' can be selected. At the first time this is done the 'New Launch Configuration Wizard' opens. From the list of different launch types the 'Application TRK Launch Configuration' is the correct choice as the application will be launched using TRK. After clicking 'Next' the proper 'TRK Connection' has to be chosen. This is the same connection that should have been already configured in the 'Connecting the Symbian device to a PC'-step.

Clicking 'Finish' will start the debugging process. The program is compiled, signed, copied to the device using TRK and started, so that after some seconds the words 'Hello World' should appear on the device's screen. Carbide will change

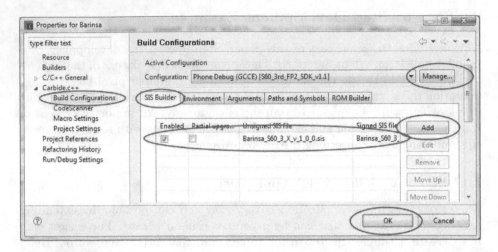

Fig. 2. Settings for different build configurations in the project's property menu

to the debug view. Views can be changed using the buttons on the upper right. Note: If during compilation an error of the kind 'HWE_0xE1382409.hlp.hrh: No such file or directory[..//src//Hweappui.cpp:21] HWE line 34 C/C++' should occur, then a wrong version of Active Perl is installed. Version 5.6.1 is required to compile the application help properly. The second launch of the application does not require all these steps. The application can be started by pressing the 'F11'-key or by using the debug button in the menu bar (see Figure 3).

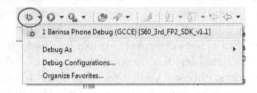

Fig. 3. Starting the debugging process using the debug button in the menu bar

4.2   Carbide Projects

For each application a so-called project is created within Carbide. A project consists of several folders and configuration files. The most important directories are 'data', 'inc', and 'src'. The 'data' directory contains resource files, such as files that contain the text for dialogs windows and menus as well as image files or the like. The 'inc' directory contains the header files of the classes as well as a '.pan' file which contains so-called 'panic constants'. A 'panic' is like a 'leave' but it can not be catched. Instead the application will terminate immediately

and return a panic error code which can be associated to a message using the
'.pan' file. Further, the 'inc' directory contains a '.hrh' file which contains several
program specific constants including the UID3.

The 'src' directory contains the program's source code in the form of '.cpp'
files. Carbide allows to quickly browse and navigate through source code files.
The 'CRTL'+'TAB' keys can be used to quickly switch between header file and
implementation file. Further, clicking any variable, class or function name enables
to jump to the definition by pressing the 'F3' key.

After creating a project from the 'S60 Gui Application' template, five files can
be found in the project's 'src' directory. If the project's name is *ProjectName*,
then the files are:

- *ProjectName*.cpp contains the start point of the application. Applications
  are loaded by Symbian OS as DLLs and they are started using the E32Main()
  method which can be found in this file. This method creates an instance of
  the C*ProjectName*Application class and returns a reference to it.
- *ProjectName*Application.cpp is the main application class file. At this
  point it has two methods, one which creates a GUI related document object
  and one which can return the application's UID3 for security checks.
- *ProjectName*Document.cpp holds the class responsible for the creation of
  the application user interface (AppUi).
- *ProjectName*AppUi.cpp holds the implementation of the application user
  interface (AppUi) class. The AppUi determines the outer appearance of the
  application and handles user input. For the display of information, an object
  of the AppView class is created by AppUi.
- *ProjectName*AppView.cpp holds the implementation of the AppView class,
  which draws the parts of the application's user interface located above the
  menu elements. In the simplest case it can just be painted white.

4.3   Using the Camera

The application that was created in the way described before simply prints text.
In the following we will describe how to extend this class in order to create a multimedia application. To do this, we extend the class C*ProjectName*AppView by
the ability to capture images from the device's built-in camera. In the following
the *ProjectName* is assumed to be CameraTest.

As described before, accessing the capabilities of a phone requires user permission. Applications that want to make use of the camera need to have the
rights to access the 'UserEnvironment' capability set in the project's mmp-file.

Within the Symbian C++ framework the CCamera-object allows to control
the camera. It allows to turn the camera on or off, select the zoom level, active
the white balance and more. The use of this object follows the event-listener
paradigm. A user application that uses the camera implements a listener interface
and will be notified of events by the CCamera object. In particular, the interface
that needs to be implemented is the MCameraObserver interface. CCamera will
call this interface's callback methods in order to deliver images.

Practically the following things need to be done.

– Add the capability 'UserEnvironment' to the mmp-file to allow the usage of the camera.
– As the CCamera class belongs to 'ecam.lib', this library has to be added to the mmp-file.
– As the CFbsBitmap class that will be used belongs to 'fbscli.lib', this library has to be added to the mmp-file too.
– Also, the class C*ProjectName*AppView (here CCameraTestAppView) has to be modified. First the changes for the header file:
  • The import of the library is necessary in order to find the CCamera object:

```
#include <ECam.h>
```

  • Under 'private' a pointer to the CCamera object and to a Bitmap that will be used later should be added:

```
CCamera* iCamera;
CFbsBitmap* iBitmap;
```

  • The class MCameraObserver has to be inherited. It only consists of virtual functions which have to be implemented. As an alternative there is MCameraObserver2 which has a simplified event handling. As this interface can be used only with S60 3rd edition, Feature Pack 2 or higher, we will use the older interface here:

```
class CCameraTestAppView : public CCoeControl,
                                    MCameraObserver
```

  • A method of MCameraObserver. It is called when CCamera::Reserve() was completed:

```
/**
 * From MCameraObserver.
 * Called when "CCamera::Reserve()" is completed.
 */
virtual void ReserveComplete(TInt aError);
```

  • A method of MCameraObserver. It is called when CCamera::PowerOn() was completed:

```
/**
 * From MCameraObserver.
 * Called when "CCamera::PowerOn()" is completed.
 */
virtual void PowerOnComplete(TInt aError);
```

  • A method of MCameraObserver which is called when a frame that was requested from the viewfinder is ready to be used:

```
/**
 * From MCameraObserver.
 * Called when "CCamera::StartViewFinderBitmapsL()"
 * is completed.
 */
virtual void ViewFinderFrameReady(CFbsBitmap& aBitmap);
```

- A method of `MCameraObserver` which is called when a still image that was requested is ready to be used:

```
/**
 * From MCameraObserver.
 * Called when "CCamera::CaptureImage()" is completed.
 */
virtual void ImageReady(CFbsBitmap* aBitmap,
                        HBufC8*     aData,
                        TInt        aError);
```

- A method of `MCameraObserver` which is called when a video stream image that was requested is ready to be used:

```
/**
 * From MCameraObserver.
 * Called when "CCamera::StartVideoCapture()"
 * is completed.
 */
virtual void FrameBufferReady(MFrameBuffer*
                                     aFrameBuffer,
                              TInt   aError);
```

Besides these changes in the header file, of course all methods must be implemented in the cpp-file, at least with an empty body.

To use pictures from the viewfinder, the following steps have to be performed:

1. Reserve camera
2. Power on camera
3. Activate viewfinder and request images
4. Wait for image ready notification
5. Copy image
6. Display image

We will show how these steps are implemented and give code examples. The following code creates the application window, creates a bitmap of size 320×240 and tries to reserve the device's camera. The bitmap size corresponds exactly the size of the viewfinder frames and the format `EColor16MU` saves each pixel of a frame in 32 bits in ARGB with 8 bits for each color channel, where the alpha channel remains unused. It is tested whether the camera is available, if not the initialization will be aborted with an error message, else a `CCamera` object is created and used to reserve the camera. Note that creating the `CCamera` instance will fail with a `KErrPermissionDenied` error if the application does not have the necessary rights.

```
#include <eikenv.h>

// Constant for the bitmap size
const TSize KBitmapSize = TSize(320,240);
[...]
```

```
void CCameraTestAppView::ConstructL(const TRect& aRect){
    // Create a window for this application view
    CreateWindowL();

    // Set the windows size
    SetRect(aRect);

    // Create bitmap
    iBitmap = new (ELeave) CFbsBitmap();
    iBitmap->Create(KBitmapSize, EColor16MU);

    if ( !CCamera::CamerasAvailable() ){
        CEikonEnv::Static()->HandleError(
                            KErrHardwareNotAvailable);
        return;
        }

    // Camera index 0 is the main camera
    iCamera = CCamera::NewL( *this, 0 );

    // Try to reserve the camera
    if (iCamera) iCamera->Reserve();

    // Activate the window, which makes it ready to be drawn
    ActivateL();
}
```

The above initialization code ends with the call of the Reserve function. No other application can use the camera after CCamera::Reserve() has been executed successfully. If the camera was successfully reserved, the CCamera object makes a callback to the ReserveComplete method where the initialization continues by turning the camera on.

```
void CCameraTestAppView::ReserveComplete(TInt aError){
    if (aError == KErrNone) iCamera->PowerOn();
}
```

If the camera was turned on successfully, a callback to PowerOnComplete is made. The next step is to activate the viewfinder. Within this step image size and other parameters can be set. The following example just sets the image size, but also white balance or zoom level can be set. The exact list of parameters that can be set can be derived by reading out TCameraInfo.

```
void CCameraTestAppView::PowerOnComplete(TInt aError){
    if (aError == KErrNone) {
        TSize size(KBitmapSize);
        TRAP_IGNORE(iCamera->StartViewFinderBitmapsL(size));
    }
}
```

The native size of the viewfinder frames may be device dependent. For the Nokia N95 phone the native size is 320×240 which will be enlarged or shrunk regarding to the target size. In any case, the aspect ratio of 4:3 is preserved and if necessary the SDK adjusts the height. In this case, native size and target size are the same so that no transformation happens. This improves speed and quality.

In this state, the CCamera object calls ViewFinderFrameReady to signalize that frames are captured and ready to be used. The frames can be copied and used further. After using the camera, the viewfinder has to be stopped, the camera has to be turned off and the reservation has to be released.

```
CCameraTestAppView::~CCameraTestAppView(){
    if (iCamera) {
        iCamera->StopViewFinder();
        iCamera->PowerOff();
        iCamera->Release();
        delete iCamera;
    }
    if (iBitmap) delete iBitmap;
}
```

We use the ViewFinderFrameReady method that is called by the CCamera object to handle the images. To copy an image it would be possible to copy the whole graphic context using memcopy. In this example we use a for loop that copies the image pixel-wise. This way it becomes clear how to access pixel data and it can be easily extended to an image filter operation.

```
void CCameraTestAppView::ViewFinderFrameReady(CFbsBitmap&
                                                        aBitmap){
    TSize iBitmapSize = aBitmap.SizeInPixels();
    TUint32* pSrc = aBitmap.DataAddress();
    TUint32* pDest = iBitmap->DataAddress();
    aBitmap.LockHeap();
    iBitmap->LockHeap();
    for(int k=0;k<iBitmapSize.iHeight*iBitmapSize.iWidth;k++)
        pDest[k]=pSrc[k];
    iBitmap->UnlockHeap();
    aBitmap.UnlockHeap();
    DrawDeferred();
}
```

The above code first determines the bitmap size and locks the heap in order to prevent the memory addresses from changing. Then, the image is copied pixel-wise and the heap is unlocked. Finally a statement recommends the redrawing of the view which will call the draw method.

```
void CCameraTestAppView::Draw(const TRect& /*aRect*/) const {
    // Get the standard graphics context
    CWindowGc& gc = SystemGc();

    // Gets the control's extent
    TRect drawRect(Rect());
```

```
// Draw Bitmap if available
if (iBitmap) {
    gc.DrawBitmap(drawRect, iBitmap, KBitmapSize);
}else{
    // Clears the screen
    gc.Clear(drawRect);
}
}
```

The modification here is just that the viewfinder frames will be rendered to the whole area which is determined by calling Rect(). This example will take the viewfinder images and display them.

Hint: If any compile errors should occur, check if 'ecam.lib' and 'fbscli.lib' have been added to the libraries in the mmp-file and that also the interface's functions ImageReady and FrameBufferReady are implemented with at least an empty body. If the program terminates directly at the beginning with the error that it has not the sufficient rights, then probably the capability 'UserEnvironment' was not added to the mmp-file.

4.4   Implementing a Pixel-Based Image Operation

Just to give a very simple example of how to access and to modify pixel data, we show how a color to gray filer is implemented. The frames are encoded as EColor16MU and the color components can be extracted using AND operations and bit shifts. The three color channels are averaged and then written back to all three color channels. This example also locks and unlocks the heap.

```
#define GetRed(c)    ((c & 0xFF0000)>>16)
#define GetGreen(c)  ((c & 0xFF00)  >> 8)
#define GetBlue(c)   (c & 0xFF)

void CCameraTestAppView::ViewFinderFrameReady(CFbsBitmap&
                                                        aBitmap){
    TInt gray;
    TSize iBitmapSize = aBitmap.SizeInPixels();
    TUint32* pSrc = aBitmap.DataAddress();
    TUint32* pDest = iBitmap->DataAddress();
    aBitmap.LockHeap();
    iBitmap->LockHeap();
    for(int k=0;k<iBitmapSize.iHeight*iBitmapSize.iWidth;k++){
        gray = (  GetRed(pSrc[k])
                + GetGreen(pSrc[k])
                + GetBlue(pSrc[k])
               )/3;
        pDest[k] = (gray<<16) | (gray<<8) | gray;
    }
    iBitmap->UnlockHeap();
    aBitmap.UnlockHeap();
```

```
DrawDeferred ();
}
```

4.5 Using Autofocus

In the following we will further extend the above example and include the use of autofocus. Concerning the autofocus it is very important to know that the way to use it has changed completely in S60 3rd Edition Feature Pack 2 compared to Feature Pack 1. The class CCamera::CCameraAdvancedSettings is new in Feature Pack 2 and allows to address autofocus and other functions. To use this the SDK has to be extended by installing a plug-in. As code generated this way will not run on Feature Pack 1 devices and because most currently existing devices including the Nokia N95 phone are Feature Pack 1 devices, we will describe the prior way to use the autofocus which will work on all devices.

In S60 3rd Edition Feature Pack 1 autofocus is NOT a part of the SDK. But Nokia provided an external library which can be used: 'S60 Platform: Camera Example with Autofocus Support'[3]. The download of this example includes the archive file 'autofocus_extension_library_s60_3rd_ed.zip' which has to be extracted to the SDK directory. To be able to use the autofocus also in debug mode, the file 'epoc32\release\armv5\urel\CamAutoFocus_s.lib' has to be copied to 'epoc32\release\armv5\udeb\'.

In the following example a slight push of the camera trigger shall start the autofocus and releasing the trigger shall stop it. To do this, a key listener will be implemented in the AppUI and the corresponding program logic will be put into the AppView.

The following things need to be done:

- The library file CamAutoFocus_s.lib has to be specified as static library within the mmp-file.
- The header *ProjectName*AppView.h of the above example has to be extended by CCamAutoFocus header.

```
#include <CCamAutoFocus.h>
```

- The main class needs to implement the interface MCamAutoFocusObserver in order to be a listener to autofocus events.

```
class CCameraTestAppView : public CCoeControl,
                           MCameraObserver,
                           MCamAutoFocusObserver
```

- Adding a method of MCamAutoFocusObserver which is called when the autofocus initialization done by CCamAutoFocus::InitL() was completed:

```
/**
 * From MCamAutoFocusObserver.
```

---

[3] http://www.forum.nokia.com/info/sw.nokia.com/id/9fd0e9a7-bb4b-489d-84ac-b19b4ae93369/S60_Platform_Camera_Example_with_AutoFocus_Support.html

```
 * Called when "CCamAutoFocus::InitL()" is completed.
 */
virtual void InitComplete(TInt aError);
```

- Adding a method of MCamAutoFocusObserver which is called when the attempt to focus ended either successfully or not:

```
/**
 * From MCamAutoFocusObserver.
 * Called when "CCamAutoFocus::AttemptOptimisedFocusL()"
 * is completed.
 */
virtual void OptimisedFocusComplete(TInt aError);
```

- Adding two variables. The first to store a reference to the autofocus object and the second to indicate active that an autofocus attempt is in progress.

```
CCamAutoFocus * iAutoFocus;
TBool iFocusing;
```

- Adding two public methods to trigger and to stop the autofocus.

```
public:
/**
 * Starts the optimized autofocus operation.
 * Does nothing if autofocus is not supported.
 */
void StartFocusL();

/**
 * Cancels an ongoing autofocus operation.
 */
void CancelFocus();
```

Besides these additions, some implementation is required. First, ConstructL will be extended by some more initialization code:

```
void CCameraTestAppView::ConstructL(const TRect& aRect){
    [...]
    // Camera index 0 is the main camera
    iCamera = CCamera::NewL( *this, 0 );

    // Try to create the autofocus object
    iFocusing = EFalse;
    TRAPD(afErr, iAutoFocus = CCamAutoFocus::NewL(iCamera));
    if (afErr)
    {
        // KErrExtensionNotSupported: deactivate autofocus
        iAutoFocus = 0;
    }
    iCamera->Reserve();
    [...]
```

```
}
```

```
// Delete object in destructor
CCameraTestAppView::~ CCameraTestAppView ()
{
    [...]
    delete iAutoFocus;
}
```

The code to initialize the autofocus is executed after the camera has been successfully activated. After executing the initialization code, InitComplete will be called. This method may contain some error handling code but is left empty for the sake of simplicity here.

```
void CCameraTestAppView::PowerOnComplete(TInt aError){
    if (aError == KErrNone){
        TSize size(KBitmapSize);
        TRAP_IGNORE(iCamera->StartViewFinderBitmapsL(size));

        // Try to initialize the autofocus control
        if (iAutoFocus){
            TRAPD(afErr, iAutoFocus->InitL(*this));
            if (afErr){
                delete iAutoFocus;
                iAutoFocus = 0;
            }
        }
    }
}

void CCameraTestAppView::InitComplete(TInt aError){
}
```

The focusing process is started when the method StartFocusL will be called. This method makes use of the variable iFocusing to check whether autofocus is already in progress. The reason to use this variable is that a second call of AttemptOptimisedFocusL will cause an 'undefined error'. So after checking availability and the variable, the focus range is set and AttemptOptimisedFocusL is called. This function calls back OptimisedFocusComplete when focusing is completed. The OptimisedFocusComplete function can be used to deal with errors or to give a feedback of success, e.g. by letting the phone vibrate, or by playing a sound. The autofocus process can be canceled. We implement the CancelFocus method to do this. This method cancels the autofocus process if it is in progress.

```
void CCameraTestAppView::StartFocusL(){
    // Do nothing if autofocus is not supported
    // or if focusing is already in progress
    if (!iAutoFocus || iFocusing)
        return;
```

```
      // Set AF range to normal before first focus attempt
      iAutoFocus->SetFocusRangeL(CCamAutoFocus::ERangeNormal);

      // Attempt focusing.
      // Calls OptimisedFocusComplete when ready.
      iFocusing = ETrue;
      iAutoFocus->AttemptOptimisedFocusL();
}

void CCameraTestAppView::OptimisedFocusComplete(TInt aError){
      iFocusing = EFalse;

      if (!aError){
          // Play a sound etc.
      }
}

void CCameraTestAppView::CancelFocus(){
      // Do nothing if autofocus is not supported
      // or if focusing is not in progress
      if (!iAutoFocus || iFocusing == EFalse)
          return;

      iFocusing = EFalse;
      iAutoFocus->Cancel();
}
```

We will call the methods StartFocusL and CancelFocus when the trigger button is slightly pressed/released. In order to do this, we implement a key handler. The header of the AppUI class has to be extended by the HandleKeyEventsL method. This method is inherited from CEikAppUi and will be called if buttons are pressed/released. Also a constant should be defined for the specific key that shall be used, in this case the trigger key.

Hint: The code of a key can be derived by using
RDebug::Printf("%d", aKeyEvent.iScanCode); which displays the key's code at runtime in the console window of Carbide.

```
/**
 * From CEikAppUi.
 * Handles key events.
 *
 * @param aKeyEvent    Event to be handled.
 * @param aType        Type of the key event.
 * @return             Response code (EKeyWasConsumed,
 *                                    EKeyWasNotConsumed).
 */
virtual TKeyResponse HandleKeyEventL(
                                const TKeyEvent& aKeyEvent,
                                TEventCode aType);
```

```
// Constant for trigger-key
const TInt KStdKeyCameraFocus = 0xE2;
```

The key-listener has to check if the key that was causing the key event was the specified one. Then, depending on whether the key was pressed or released either StartFocusL or CancelFocus will be called.

```
TKeyResponse CCameraTestAppUi::HandleKeyEventL(
                               const TKeyEvent& aKeyEvent,

                               TEventCode aType){
    // Handle autofocus request
    if (aKeyEvent.iScanCode == KStdKeyCameraFocus){
        // When the focus key is pressed down,
        // start the focus operation.
        // When it is released,
        // cancel the ongoing focus operation.
        switch (aType){
        case EEventKeyDown:
            iAppView->StartFocusL();
            break;
        case EEventKeyUp:
            iAppView->CancelFocus();
            break;
        default:
            break;
        }
        // Inform system that key event was handled
        return EKeyWasConsumed;
    }
    // Inform system that key event was not handled
    return EKeyWasNotConsumed;
}
```

At this point we have completed a mobile application that uses autofocus and applies a pixel-based image operation.

4.6 Tips and Improvements

This example's code was kept as simple as possible and is not meant as an example for good software engineering. In the following we want to give some tips for improvement.

- A professional implementation should contain more error handling code.
- It should be considered to implement a separate class for the autofocus and camera control.
- In order to make applications ready to use on different devices it is a good advice to read the camera properties of a given device using

```
TCameraInfo camInfo;
iCamera->CameraInfo(camInfo);
```

- We used the easy draw method to show the image which is a little bit slow. Using gc.BitBlt is faster.
- For the actual image manipulation algorithm, a separate class should be used.
- If image filter operations are used, the filters should be separated for speed reasons, if possible.
- Due to the caching behavior images should be iterated in X-direction. Iterations in Y-direction will lead to more cache misses.
- Processors are a lot faster using integers compared to float values, so integers should be used wherever possible.
- On the N95 and other devices with a 32-bit processor, using 1-byte data types compute slightly slower than 4-byte data types.
- For the debugging break points can be used. Moreover, the static RDebug class should be used which can create output within Carbide at runtime.
- Imaging applications suffer from the behavior of the display which automatically goes off after some time without user-interaction. To prevent this from happening, the function User::ResetInactivityTime() can be used.

Dealing with camera facilities is not a straightforward undertaking. In addition to the method described in this tutorial, there is an alternative for more recent smart phones based on S60 3rd Edition and newer versions. To provide developers an easy-to-use interface for camera functions, Forum Nokia has developed a CameraWrapper [5]. This wrapper provides a unified interface for various Symbian and S60 camera APIs, some of which have previously been Feature Pack specific or only available via an SDK plug-in. The supported features include autofocus, digital zoom, flash, and exposure modes.

5 Summary and Conclusion

In this paper we have shown how to set up the hardware and the integrated software development environment for the implementation of applications for Symbian OS devices. We have explained the software development process, coding conventions and mechanisms such as exception handling and safe object creation. In the main part of this paper the step-by-step development of an application was explained which makes use of the device's camera, the autofocus functionality and which applies a basic pixel-based image operation.

From our own experiences we can say that sometimes software development using Symbian C++ can be frustrating. The reason is that sometimes features are not well documented and tutorials are missing. With this paper we want to close this gap and provide the important information that we would have needed at the beginning. We hope to have created a valuable document that will help developers who are new to Symbian C++ to save time.

References

1. Ahmad, K.: About Symbian UIDs,
   http://www.newlc.com/en/about-symbian-uids
2. Borges, R.: LEAVE and TRAP,
   http://www.newlc.com/en/LEAVE-and-TRAP.html
3. Bruns, E., Brombach, B., Bimber, O.: Mobile phone-enabled museum guidance with adaptive classification. IEEE Computer Graphics and Applications 28(4), 98–102 (2008)
4. Chang, G., Tan, C., Li, G., Zhu, C.: Developing mobile applications on the Android platform. In: Jiang, X., Ma, M.Y., Chen, C.W. (eds.) Mobile Multimedia Processing: Fundamentals, Methods, and Applications. Springer, Heidelberg (2010)
5. CameraWrapper,
   http://wiki.forum.nokia.com/index.php/
   TSS001129_-_Using_the_camera_autofocus_feature_in_S60_devices
6. Nokia: Carbide.c++ Version 2.0,
   http://www.forum.nokia.com/Tools_Docs_and_Code/Tools/IDEs/Carbide.c++/
7. Harrison, R.: Symbian OS C++ for Mobile Phones, vol. 1. John Wiley & Sons, Chichester (2003)
8. Harrison, R.: Symbian OS C++ for Mobile Phones: Programming with Extended Functionality and Advanced Features, vol. 2. John Wiley & Sons, Chichester (2004)
9. Harrison, R., Shackman, M.: Symbian OS C++ for Mobile Phones, vol. 3. John Wiley & Sons, Chichester (2007)
10. Ta, D.-N., Chen, W.-C., Gelfand, N., Pulli, K.: SURFTrac: Efficient tracking and continuous object recognition using local feature descriptors. In: Proc. of CVPR 2009, pp. 2937–2944 (2009)
11. Rome, A., Wilcox, M.: Multimedia on Symbian OS: Inside the Convergence Device. John Wiley & Sons, Chichester (2008)
12. Rothaus, K., Roters, J., Jiang, X.: Localization of pedestrian lights on mobile devices. In: Proc. of APSIPA Annual Summit and Conference, Sapporo, Japan (2009)
13. Shetty, G.: Exception handling and cleanup mechanism in Symbian,
    http://www.newlc.com/Exception-Handling-and-Cleanup.html
14. Strumillo, P., Skulimowski, P., Polanczyk, M.: Programming Symbian smartphones for the blind and visually impaired. In: Kacki, E., Rudnicki, M., Stempczynska, J. (eds.) Computers in Medical Activity, pp. 129–136. Springer, Heidelberg (2009)
15. Symbian Developer Network: An Introduction to R Classes,
    http://developer.symbian.com/main/downloads/papers/R_Classes_v1.0.pdf
16. Symbian Foundation Limited: Complete Guide To Symbian Signed,
    http://developer.symbian.org/wiki/index.php/
    Complete_Guide_To_Symbian_Signed
17. Symbian Foundation Limited: Symbian Signed,
    https://www.symbiansigned.com/app/page
18. Wachenfeld, S., Terlunen, S., Jiang, X.: Robust 1-D barcode recognition on camera phones and mobile product information display. In: Jiang, X., Ma, M.Y., Chen, C.W. (eds.) Mobile Multimedia Processing: Fundamentals, Methods, and Applications. Springer, Heidelberg (2010)

Developing Mobile Applications on the Android Platform

Guiran Chang[1], Chunguang Tan[1], Guanhua Li[1], and Chuan Zhu[2]

[1] Northeastern University, Shenyang, China
chang@neu.edu.cn, tcg1978@gmail.com, liguanhua2724@126.com
[2] Hohai University, Changzhou, China
zhu.ca@hotmail.com

Abstract. Android is a new mobile platform. Developments of mobile applications on Android have attracted a lot of attention and interest in research and industry communities. Android is the first free, open source, and fully customizable mobile platform. In this chapter, we walk through steps in developing a mobile application on the Android platform. Through an exemplary application of EPG (Electronic Program Guide) recommender framework, we present key steps in developing an Android application including how to create Android project and class, how to build simple user interface, how to utilize networking protocols and how to handle events. With the increased attention of Android and growing number of Android based mobile phones available on the market, we believe this chapter will timely help researchers rapidly prototype mobile applications to prove concepts of their research results.

Keywords: Android platform, Mobile TV, EPG recommender, Java.

1 Introduction

With mobile devices becoming more ubiquitous, powerful new tools are necessary to enable mobile businesses with faster time to market development. Android has the potential of bringing about a long-needed change to the industry [8]. Android is the first free, open source, and fully customizable mobile platform, a software stack for mobile devices including an operating system, middleware and key mobile applications. The Android SDK provides the tools and APIs necessary to develop applications on the Android platform [13][14].

EPG (Electronic Program Guide) is often a critical part of TV programming as it provides the scheduling of TV programs. Digital EPG is often provided by the MSO (Multi-Service Operator) as part of subscription and can be used by end users to browse, select and record programs. A hybrid EPG recommender architecture was presented in our prior work [9][10][21]. In this chapter, we walk through key steps of developing the EPG recommender on the Android platform with the aim of covering major functions of Android SDK in building a typical Android application.

The remainder of this chapter is organized as follows. Section 2 gives a brief introduction to the Android platform. Section 3 describes the hybrid EPG recommender system architecture and implementation issues. Section 4 presents our

X. Jiang, M.Y. Ma, and C.W. Chen (Eds.): WMMP 2008, LNCS 5960, pp. 264–286, 2010.

experience with the development of an EPG recommender core on Android. Section 5 discusses other development issues. A concluding remark is given in Section 6.

2   Getting Started with Android

2.1   The Android Platform

Android is a software platform and operating system for mobile devices based on a Linux kernel, it was initially developed by Google and later under the Open Handset Alliance. The platform is based on the Linux 2.6 kernel and comprises an operating system, middleware stack, customizable user interface and applications. It allows developers to write managed code in Java, controlling the device via Google-developed Java libraries. Applications written in C/C++ and other languages can be compiled to ARM native code and run, but this development path is not officially supported by Google [14][19].

Fig. 1 shows the major components of the Android system architecture [14]. At the bottom layer, Android relies on Linux version 2.6 for core system services such as security, memory management, process management, network stack, and driver model.

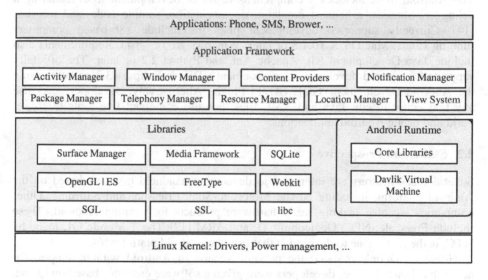

Fig. 1. The Android system architecture

Android includes a set of core libraries that provide most of the functionality available in the core libraries of Java. Every Android application runs in its own process, with its own instance of the Dalvik virtual machine. The Dalvik VM executes files in the Dalvik Executable (.dex) format which is optimized for minimal memory footprint [12].

Android includes a set of C/C++ libraries used by various components of the Android system. These capabilities are exposed to developers through the Android application framework. Android is also shipped with a set of core applications including an email client, SMS program, browser, calendar, maps, contacts, and others. All applications are written in Java. Developers have full access to the same framework APIs used by the core applications. The application architecture is designed to simplify the reuse of components. Any application can publish its capabilities and any other application may then make use of those capabilities. This same mechanism allows components to be replaced by the user. Underlying all applications is a set of services and systems, including:

- A rich and extensible set of *Views* that can be used to build an application, including lists, grids, text boxes, buttons, and an embeddable web browser;
- Content Providers that enable applications to access data from other applications or to share their own data;
- A Resource Manager that provides access to non-code resources such as localized strings, graphics, and layout files;
- A Notification Manager that enables all applications to display custom alerts in the status bar;
- An Activity Manager that manages the life cycle of applications and provides a common navigation backstack.

The Android SDK includes a comprehensive set of development tools including a debugger, libraries, a handset emulator, documentation, sample code, and tutorials [14]. Currently supported development platforms include x86-based computers running Linux, Mac OS X 10.4.8 or later, Windows XP or Vista. Requirements also include Java Development Kit, Apache Ant, and Python 2.2 or later. The officially supported integrated development environment (IDE) is Eclipse (3.2 or later) using the Android Development Tools (ADT) Plugin. Alternatively, developers may use any text editor to edit Java and XML files then use command line tools to create, build and debug Android applications.

2.2 Status and Prospective

Several manufacturers are interested in developing handheld products based on the Android platform, including Nokia, Motorola, Sony Ericsson, and Samsung. Other companies are also making ready hardware products for running Android. These include Freescale, NEC, Qualcomm, TI, and ARM [17]. The T-Mobile G1, made by HTC, is the first phone to the market that uses the Android platform [6].

However, Google released the preview version of Android without support for actual hardware. Instead, developers were given a software emulator based on Qemu. Google released the SDK with a few demonstration applications and is relying on third parties to come up with the rest [3]. The early feedback on developing applications for the Android platform was mixed [15]. Issues cited include bugs, lack of documentation, inadequate QA infrastructure, and no public issue-tracking system. Despite this, Android-targeted applications began to appear in a short time after the platform was announced.

One of the problems with Android is that software installed by users must be written in Java. This provides end-users with less control over their phone's functionality [2]. Another issue is related to Android's disregard of established Java standards, i.e. Java SE and ME. This prevents compatibility among Java applications written for those platforms and those for the Android platform. Android only reuses the Java language syntax, but does not provide the full-class libraries and APIs bundled with Java SE or ME. Instead, it uses the Apache Harmony Java implementation [12].

The ADB (Android Debug Bridge) debugger gives a root shell under the Android Emulator which allows native ARM code to be uploaded and executed. ARM code can be compiled using GCC on a standard PC. Running native code is complicated by the fact that Android uses a non-standard C library (known as Bionic). The underlying graphics device is available as a framebuffer at */dev/graphics/fb0* [4]. The graphics library that Android uses to arbitrate and control access to this device is called the Skia Graphics Library (SGL) [18].

Native classes can be called from Java code running under the Dalvik VM using the *System.loadLibrary* call, which is part of the standard Android Java classes [14]. Elements Interactive B.V. have ported their Edgelib C++ library to Android, and native code executables of their S-Tris2 game and Animate3D technology demo are available for download [19]. Native code can be executed using the ADB debugger, which is run as a background daemon on the T-Mobile G1 [7].

Android started to enter the market for mobile phone software when it was already crowded. However, Android had aroused great enthusiasm [17] and its market share had been growing rapidly. In Mid October 2008, it was reported that the pre-sale of T-Mobile G1 had hit 1.5 million [11] and new Android-powered devices were coming to market or under development [1][6].

The close relation of Android with Linux is likely to encourage the development of a wave of new handset applications. A lot more software will be seen to run on devices from people who want to exploit the commonality with desktop systems [17].

2.3 Install and Configure the Android SDK and Eclipse Plug-In

To develop Android applications using the code and tools in the Android SDK, the operating systems could be: Windows XP or Vista, Mac OS X 10.4.8 or later (x86 only), and Linux. The supported development environments are: Eclipse IDE and other development environments or IDEs [14].

- Eclipse IDE: Eclipse 3.3 (Europa), 3.4 (Ganymede); JDK 5 or JDK 6 (JRE alone is not sufficient); Android Development Tools plugin (optional).
- Other development environments or IDEs: JDK 5 or JDK 6 (JRE alone is not sufficient); Apache Ant 1.6.5 or later for Linux and Mac, 1.7 or later for Windows.

The recommended way to develop an Android application is to use Eclipse with the ADT plugin. This plugin provides editing, building, and debugging functionality integrated right into the IDE. However, applications can also be developed in another IDE, such as IntelliJ, or use Eclipse without the ADT plugin. The steps for installing and configuring the Android SDK and Eclipse Plug-in are as follows.

1) Download the SDK (http://code.google.com/intl/en/android/download.html).
2) Install the SDK: After downloading the SDK, unpack the .zip archive to a suitable location on the computer. By default, the SDK files are unpacked into a directory

named android_sdk_*<platform>*_*<release>*_*<build>*. The directory contains the subdirectories tools/, samples/, and others. Adding *tools* to the path allows running Android Debug Bridge (ADB) and the other command line tools without needing to supply the full path to the tools directory.

3) Install the Eclipse plugin (ADT): To use the Eclipse IDE as the environment for developing Android applications, a custom plugin called Android Development Tools (ADT) is to be installed, which adds integrated support for Android projects and tools. The ADT plugin includes a variety of powerful extensions that make creating, running, and debugging Android applications faster and easier. If the Eclipse IDE is not used, then the ADT plugin needs not to be downloaded and installed.

3   The EPG Recommender System on Mobile Devices

In recent years, many different mobile TV standards and products are emerging and electronic programming guide (EPG) is provided to consumers on mobile devices. As the number of channels available on the existing and future TV broadcasting network increases, an EPG recommender system becomes eminent in meeting the needs of delivering a prioritized listing of programs. With the convergence of TV broadcasting for mobile device and cable network at home, the EPG recommender system has to adapt itself to a mobile environment [9][10].

In this chapter, an EPG recommender is used as an example for building the Android application. A simplified architecture of the EPG recommender [10] is shown in Fig. 2 to give readers needed pre-requisite. The details of the EPG recommender framework can be found in our companion paper [10]. The EPG recommender is implemented as an Android application which generates a prioritized listing of programs for the user based on the user profile and the EPG database.

3.1   Building Blocks of the EPG Recommender on a Mobile Device

In Fig. 2, the left part shows the components of the EPG recommender on the mobile device and the right part shows those on the home server. The hybrid architecture supports a stand-alone mobile EPG application and a client/server EPG application. In the stand-alone application, an EPG service middleware with simple built-in recommender is implemented and fully functioning on the mobile device. In the client/server application, the home server provides full EPG recommender service and the mobile device is acting as a browser in a slave mode.

There are four filters in the simple recommender module for the mobile device, including time, station, category, and domain filters. The other major components of the EPG application include a friendly user interface, a user profile, the handoff operations, the EPG database and its related operations.

A simplified user profile residing on the mobile device only keeps the latest user preferences. For the implementation on Android, a single user profile is required to fit into two forms of EPG: one from DVB-H (Digital Video Broadcast - Handheld) on the mobile device and the other from traditional broadcast or Internet on the server.

An EPG database is to be constructed which is simple and suitable to the recommender on the mobile.

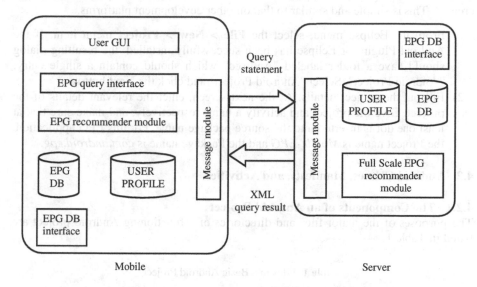

Mobile   Server

Fig. 2. A hybrid EPG recommender and its network architecture

3.2 Functions Implemented on the Android SDK

A prototype of the simple EPG recommender system is developed on the Android platform with UDP adopted as the network protocol for communication between the mobile device and the server. A laptop computer is used to run the Android SDK to simulate a mobile device. The recommendation filters have been implemented but will not be discussed in more detail. In the next section, our development experience on the Android is presented.

4   Developing the EPG Recommender Core on Android

To develop an application on Android, the developer needs to understand Java. Some familiarity with other mobile platforms will be helpful. Android's activity is quite similar to BREW's applet [16] or Java ME's midlet. The mobile UI development on Android is like a happy union of Java ME's Canvas/Screen object and the BREW widget hierarchy [8].

The process of learning how to build applications for a new API is pretty similar. Generally, there are two phases: first, learn how to use the APIs to do what is to be done; then learn the nuances of the platform. The second phase is to learn the right way to build applications. Successful applications will offer an outstanding end-user experience [14].

4.1 Create an Android Project

Before developing a new application on Android, an Android project has to be created. This is simple and similar to that on other development platforms.

1) From the Eclipse menu, select the **File > New > Project** menu item. If the Android Plugin for Eclipse has been successfully installed, the resulting dialog should have a folder labeled "Android" which should contain a single entry: "Android Project". Select "Android Project" and click the **Next** button.
2) Fill out the project details. At the next screen, enter the relevant details of the project. Give the project and activity a name respectively, and make sure that at least one dot is inserted into the source package name. For this EPG application, the Project name is *Mobile-EPG* and the Package name is *com.android.epg*.

4.2 Android Classes, Manifests, and Activities

4.2.1 The Components of an Android Project

The purposes of the major files and directories of a functioning Android project are listed in Table 1.

Table 1. Files in a Basic Android Project

FILE NAME	PURPOSE
YourActivity.java	File for default launch activity; more on this to follow.
R.java	File containing an ID for all asset constants.
Android Library/	Folder containing all Android's SDK files.
assets/	Multimedia and other miscellaneous required files.
res/	Base directory for resources used by the UI.
res/drawable	Directory for image files to be rendered by the UI layer.
res/layout	All XML-style view layout files are stored here. Again, more on this later.
res/values	Location for string's and configuration files.
AndroidManifest.xml	File that describes your application to the outside operating system.

Among various files the most important one is the Android manifest. It is the link between the EPG application and the outside world. In this file, the intents for activities will be registered. So this will be the groundwork for further development. Code Listing 1 shows the Android manifest for the EPG application.

After the xml/declaration comes the *<manifest>* tag with the schema URL. One of its properties is the name of the source package, *com.android.epg*. Next is the application declaration and the location of the application icon. The *@drawable* notation indicates that the item can be found in the res/drawable folder. The EPG application consists of a single activity: *.mobile_epg*.

The activity name is hard-coded, but its label, or what is displayed to the user within the Android application menu, is defined in the res/values/strings.xml file. The *@string* notation pulls the value from *app_name* within the aforementioned XML file.

android.intent.action.MAIN and *android.intent.category.LAUNCHER* in the intent filter description are predefined strings that tell Android which activity to fire up as the EPG application is started.

The next in the important file list is R.java which contains the reference identification numbers for all the resources, both graphical and drawable. Adding a new graphic to res/drawable will result in a corresponding identifier in the *R.java* class. References can then be made to these IDs rather than the items in the file system. This allows swapping out strings for localization, screen size, and other errata that are guaranteed to change between devices.

Code Listing 1. Android Manifest.xml

```xml
<?xml version="1.0" encoding="utf-8"?>
<manifest xmlns:android="http://schemas.android.com/apk/res/android"
        package="com.android.epg">
    <application android:icon="@drawable/icon" android:label=
            "@string/app_name">
        <activity android:name=".mobile_epg" android:label=
                "@string/app_name">
            <intent-filter>
                <action android:name="android.intent.action.MAIN" />
                <category android:name=
                        "android.intent.category.LAUNCHER" />
            </intent-filter>
        </activity>
    </application>
</manifest>
```

4.2.2 Important Activities and Classes in the EPG Application

An application in Android is defined by the contents of its manifest. Each Android application declares all its activities, entry points, communication layers, permissions, and intents through AndroidManifest.xml. Four basic building blocks are combined to comprise an Android application:

1) Activity: The most basic building block of an Android application,
2) Intent receiver: A reactive object launched to handle a specific task,
3) Service: A background process with no user interface,
4) Content provider: A basic superclass framework for handling and storing data.

Except the activity block, other three blocks are not needed in all applications. The EPG application will be written with some combination of them. There are two packages in directory [src/], com.android.epg and com.android.net. There are seven java files in the first package: Mobile_EPG.java, NotesDbAdapter.java, publicdata.java, R.java, set_tab.java. splashyframe.java, and tab_config.java. In the second package, there are two java files, Network.java and UdpClt.java. A brief description to these files is as follows.

R.java is an auto-generated file. This class is automatically generated by the AAPT (Android Asset Packaging Tool) from the resource data it finds. It cannot be modified and should not be modified by hand.

The *Splashyframe* class is used to display the initial interface of EPG application which shows some information about this application, e.g., the logo and the version number. This interface will disappear after 3000ms and then the EPG application will enter the configuration interface. User can hit the w key within 3000ms to enter the configuration interface earlier.

Set_tab extends *TabActivity*. *TabActivity* is an activity that contains and runs multiple embedded activities or *Views*. Through this activity, the application will get three tabs named Browse, Recommend and Configure, respectively.

The major function of *tab_config* activity is to use *setContentView*(R.layout.config) function to generate an UI for configuring the user information. This activity sets the activity content from a layout resource. The resource will be inflated, adding all top-level *Views* to the activity. *Spinner* is one of these top-level *Views*, which displays one child at a time and lets the user pick among them. The items in the *Spinner* come from the *Adapter* associated with this *View*. An *Adapter* object acts as a bridge between an *AdapterView* and the underlying data for that *View*. The *Adapter* provides access to the data items. The *Adapter* is also responsible for making a *View* for each item in the data set, see Code Listing 2.

Code Listing 2. Setting ArrayAdapter

```
s1=(Spinner) findViewById(R.id.spinner4);
 ArrayAdapter<CharSequence> adapter=ArrayAdapter.createFromResource
     (this, R.array.config, android.R.layout.simple_spinner_item);
adapter.setDropDownViewResource(android.R.layout.
     simple_spinner_dropdown_item);
s1.setAdapter(adapter);
```

The *Mobile_EPG* activity uses the function *setContentView*(R.layout.list1) to display a user interface. This UI is the main query interface of the EPG application and the user can select the category, station, time range and date of the interested program. The function *onClick* in *Button.OnClickListener* will be called when the *View* has been clicked (see Code Listing 3). If the status is HOME, the EPG application will connect to the server with the given IP address and port number in activity *tab_config* and send the SQL statement to the server (see Code Listing 5). The server will send the result set to the handset in XML format and then the application will resolve the XML string with the function *parserXMLString* (see Code Listing 4) and use the function *fillHomeData* (see Code Listing 6) to display the result set which is received from the server. If the status is REMOTE, the EPG application will connect to the local database with query condition (the operations on database will be explained in Section 4.5) and use function *fillRemoteData* (see Code Listing 7) to display the result set obtained from the local database. If the status is OFFLINE, the application will inflate a dialog with the message "sorry, the status is offline!". Code Listing 8 shows how to display the message on the Android screen.

Code Listing 3. Obtain the mobile status

```
 private Button.OnClickListener dis_list = new
     Button.OnClickListener(){
   public void onClick(View v) {
       if (publicdata.status.equalsIgnoreCase(Mobile_EPG.status_home))
          {...}
       else if(publicdata.status.equalsIgnoreCase
          (Mobile_EPG.status_remote))
          {...}
       else {...}}}
```

Code Listing 4. ParserXML

```
qrs=udpclt. sentMsg
       (condition);
mStrings=parserXMLString
       (qrs);
```

Code Listing 5. NetWork

```
UdpClt udpclt=new UdpClt();
UdpClt.setSrvAddr(publicdata.ip);
UdpClt.setSrvPort(Integer.parseInt
       (publicdata.port));
... ...
qrs=udpclt.sentMsg(condition);
```

Code Listing 6. fillHomeData

```
private void fillHomeData(){
    setListAdapter(new ArrayAdapter<String>(this,
           android.R.layout.simple_list_item_1, mStrings));
    getListView().setTextFilterEnabled(true);}
```

Code Listing 7. fillRemoteData

```
private void fillRemoteData() {
    mNotesCursor=mDbHelper.fetchAllNotes (sql_station, sql_date,
        sql_time_range, sql_category);
    startManagingCursor(mNotesCursor);
    String[] from = new String[] {NotesDbAdapter.KEY_STATION,
        NotesDbAdapter.KEY_TIME, NotesDbAdapter.KEY_TITLE};
    int[] to = new int[] { R.id.text1, R.id.text2, R.id.text3};
```

Code Listing 8. Display a message on Android screen

```
openErrorDialog() {
    new AlertDialog.Builder(this)
    .setTitle(R.string.error)
    .setMessage(dialog_error_msg)
    .setNegativeButton(R.string.homepage_label, new DialogInterface.
           OnClickListener() {
       public void onClick(DialogInterface dialoginterface,
              int i){}})
    .show();}
```

UdpClt is a normal java class built in package *com.android.net* which can send massages to the server with the UDP protocol. The address of the server can be obtained from the GUI configuration. When using UDP transmission, the client just waits for a fixed time interval. If no response is received, it means that the server is not available. Normally, the *socket.receive(packet)* function will wait for ever, until it gets the response. Thus, a timer is set (such as 500 milliseconds). If time out occurs, the client will choose mobile local built-in recommender engine. In case that the action for storing response of recommendation to packet.data is interrupted by the time out process on the client, the *INTEGRALITYFLAG* is used to indicate the integrality of the response. The client checks the packet received from the server to determine whether the packet is ended with *INTEGRALITYFLAG* (see Code Listing 9).

Code Listing 9. Checking integrality of the response

```
if(response.length()>=INTEGRALITY_FLAG.length()&&(response.substring
    ((response.length()-10),  response.length()).equalsIgnoreCase
    (INTEGRALITY_FLAG))) {
    System.out.println(response.substring(0, (response.length()-
        INTEGRALITY_FLAG.length())));
    socket.close();
```

```
         return response.substring(0, (response.length()-INTEGRALITY_FLAG.
             length())); }
else {return LINK_UNSTABLE; }
```

4.3 Construct a User Interface

4.3.1 Development of a Simple Android User Interface

After running the Android emulator, we will see a text field with "Hello World, mobile_epg" in the Android emulator window. This is adapted from the default user interface example of Android procedure architecture and is implemented by the code in the file res/layout/main.xml in the IDE window as shown in Fig. 3.

The string "*Hello World, mobile_epg*" is in line 10 of Fig.3. In the Android platform, the user interface is displayed by the class *ViewGroup* or *View*. *ViewGroup* and *View* are the basic expression elements of the user interface in the Android platform and are called in our EPG application. This means that the interface elements used to display on the screen and the application logic are mixed together. We can also separate the interface elements from the application logic to use XML format file to describe the organization of the interface components in accordance with the clear way provided by Android. The EPG application is developed with the second approach. The main.xml document is described as follows.

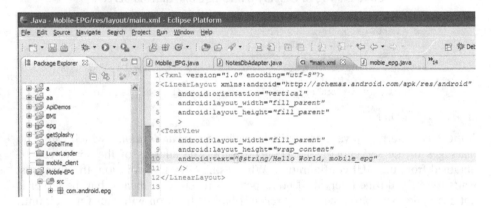

Fig. 3. The file res/layout/main.xml in the IDE window

The first line in file main.xml is the fixed content of XML file. In the next line, the label *LinearLayout* is used to express the region of the interface elements. *LinearLayout* organizes its children into a single horizontal or vertical row. It creates a scrollbar if the length of the window exceeds the length of the screen. The string "*android*" after xmlns is used to declare the namespace of the XML document. The following URL is used to express that the document will refer to the definition of Android namespace. This attribute must be included in the label of Android Layout Document.

From line 3 to 5, the narratives included in label *LinearLayout are its* attributes. Most of the attributes of the Android application labels in the directory *layout* have a prefix *"android:"*. The interface elements have many attributes in common, e.g. length, width. The length and width of Android interface element are set by *"android:layout_width"* and *"android:layout_height"*, respectively. The parameter value *"fill_parent"* means filling the entire screen, or we can set specific values (such as 20ps, 35pix) for the attributes. We set *android:orientation* to *vertical,* which means it aligns all its children in a single direction one by one.

Android.view.ViewGroup is the basic class of many layouts and *Views.* The common implementions are: *LinearLayout, FrameLayout, TableLayout, AbsoluteLayout,* and *RelativeLayout.* There is no need to change the default *LinearLayout* for most applications. What we have to do is filling in the required interface components in the *LinearLayout.* Thus the content after line 7 will normally be modified when a common Android application is developed. *TextView* is the first familiar interface component to the user which displays text and optionally allows the user to edit it.

In line 8 to 10 are the attributes contained in *TextView.* We set *android:layout_width* to *fill_parent,* which means that the width will be the same as the *parent* in *LinearLayout.* A new attribute *wrap_content* is used in *android:layout_height* which means that the hight of the interface component will change with the difference in the number of lines. The attribute *android:text* is the content to be displayed on the component.

4.3.2 Android Layout of the EPG Application
There are five XML files in the EPG application directory *res/layout/.* They are alert_dialog_text_entry.xml, config.xml, list1.xml, notes_row.xml and splash.xml.

The splash.xml file defines the booting interface of the EPG application for a better user experience as shown in Code Listing 10. The interface element *imageview* has an attribute *android:src* with value *@drawable/menu_background*. It means that the picture shown in the screen (see Fig. 4) is just stored in the file with the name menu_background.jpg in the directory *res/drawable/.*

In the user configuration interface shown in Code Listing 11, the *ipaddress* and *portnumber* field is used to set the IP address and port number of the server in HOME status. The *delay* field is the time delay the EPG user can accept. The *network status* field is used to select the current status of the mobile, HOME, REMOTE, or OFFLINE. The user configuration takes effect only when the button *Set-Configuration* is clicked. The value of the attribute of *EditText, android:numeric,* is set to be *integer,* which requires the EPG user to input only integer figures for *EditText.* Its attribute *android:text* is set to the default value. Next is the button component which has two attributes *android:layout_width* and *android:layout_height.* The other attribute, *android:text,* is used to display the characters on the button. The *Spinner* widget is a *View* that displays one child at a time and lets the user pick among them. The items in the *Spinner* come from the *adapter* associated with this *View.* An *Adapter* object acts as a bridge between an *AdapterView* and the underlying data for that *View.* The XML attribute *android:prompt* is the prompt to display when the *Spinner's* dialog box is shown. We will see that box (see Fig. 5 (b)) after clicking the arrowhead at the right side in Fig. 5 (a). An *Adapter* is created with the given array and made to be associated with the *Spinner* as shown in Code Listing 13.

Code Listing 10. splash.xml

```xml
<?xml version="1.0" encoding="utf-8"?>
<LinearLayout xmlns:android=
      "http://schemas.android.com/
      apk/res/android"
   android:orientation="vertical"
   android:layout_width="fill_parent"
   android:layout_height="fill_parent">
   <ImageView android:src="@drawable/
      menu_background"
    android:layout_width="fill_parent"
    android:layout_height="fill_parent">
   </ImageView>
</LinearLayout>
```

Fig. 4. The booting interface of EPG

The user selects the conditions of the program requested by using the user interface with the browse tab (Fig. 6). If there are programs that satisfy the conditions, the result list will be displayed here. There are three *Spinners* in this interface with the default value *All Station*, *All Category*, and *All Day* (see Fig. 6 (a)). They are used to select the station, category, and time range of the requested program for the user. There are also two *Button* widgets. The left button, whose default value is current date on the screen, is used to select the date of the program. After it is clicked, the interface will be that shown in Fig. 6 (b). After all the conditions are chosen, the right button, named "*ok, go*", is clicked and then the background procedure catches the *click* event and connects to the database to search for the program. At last, the program list will be displayed on the *ListView* area as shown in Fig. 6 (c). The codes for this interface are in Code listing 12. Like the interface component Spinner, the data to be display on *ListView* comes from *ArrayAdapter. Mstrings* is the string array that will be listed on the *ListView* as shown in Code Listing 13. The event handling and operations on database will be introduced in Sections 4.4 and 4.5.

4.4 Android Event Handling Mechanism

We have demonstrated how to use Android's *TextView*, *Button*, and *EditText* widgets. Now the *text-entry* field is to be analyzed to ingest what the user has entered. Now we make *button_ok* defined in config.xml as an example to illuminate it. To access the contents of the *EditText* widgets defined earlier, the following two steps are needed: get an object handle to the widget defined in the XML; listen for clicks or select events on the *button_ok* widget.

The first task is to get pointers to the elements defined in the XML layout files. To do this, make sure that each XML widget to be accessed has an *android:id* parameter. As mentioned earlier, using the notation @+id/id_name_here will make sure the R.java file has the ID needed. The following is an example about how to get a pointer to that *View* object when the application starts up (see Code listing 14). This is the *onCreate* method in the *Mobile_EPG* activity.

Code Listing 11. config.xml

```xml
<?xml version="1.0" encoding="utf-8"?>
<LinearLayout xmlns:android="http://
    schemas.android.com/apk/res/
    android"
    android:orientation="vertical"
    android:layout_width="fill_parent"
    android:layout_height="fill_parent" >
    <TextView android:layout_width=
      "fill_parent"

android:layout_height="wrap_content"
    android:text="ipaddress:" />
    <EditText android:id="@+id/ip"
      android:layout_width="fill_parent"

android:layout_height="wrap_content"
    android:text="" />
    <TextView android:layout_width=
      "fill_parent"

android:layout_height="wrap_content"
    android:text="portnumber:" />
    <EditText android:id="@+id/port"
      android:layout_width="fill_parent"

android:layout_height="wrap_content"
    android:numeric="integer"
    android:text="6543" />
    <TextView android:layout_width=
      "fill_parent"

android:layout_height="wrap_content"
    android:text="delay:" />
    <EditText android:id="@+id/weight"
      android:layout_width="fill_parent"

android:layout_height="wrap_content"
    android:numeric="integer"
    android:text="500ms" />
    <TextView android:layout_width=
      "fill_parent"

android:layout_height="wrap_content"
    android:text="network status:" />
    <Spinner android:id="@+id/spinner4"
      android:layout_alignParentLeft=
        "true"
      android:layout_width="fill_parent"

android:layout_height="wrap_content"
    android:drawSelectorOnTop="true"
    android:prompt="@string/
      spinner_4_prompt" />
    <Button android:id="@+id/set_config"
      android:layout_width="wrap_content"

android:layout_height="wrap_content"
    android:text="Set-Configuration"/>
</LinearLayout>
```

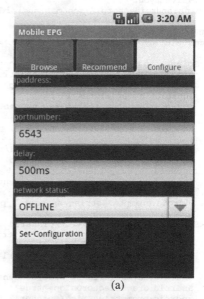

(a)

(b)

Fig. 5. (a) User configuration interface. (b) The Spinner's prompt.

Code Listing 12. list1.xml

```xml
<?xml version="1.0" encoding="utf-8"?>

<RelativeLayout
  xmlns:android="http://schemas.android.com/
     apk/res/android"
  android:layout_width="fill_parent"
  android:layout_height="wrap_content">

  <Spinner android:id="@+id/spinner1"
    android:layout_alignParentLeft="true"
    android:layout_marginRight="10dip"
    android:layout_width="155px"
    android:layout_height="wrap_content"
    android:drawSelectorOnTop="true"
    android:prompt="@string/spinner_1_prompt" />

  <android:layout_alignBaseline="@+id/spinner2"
    android:layout_alignBaseline="@id/spinner1"
    android:layout_width="155px"
    android:layout_height="wrap_content"
    android:drawSelectorOnTop="true"
    android:prompt="@string/spinner_2_prompt" />

  <Spinner android:id="@+id/spinner3"
    android:layout_alignParentLeft="true"
    android:layout_marginLeft="0dip"
    android:layout_below="@id/spinner1"
    android:layout_width="155px"
    android:layout_height="wrap_content"
    android:drawSelectorOnTop="true"
    android:prompt="@string/spinner_3_prompt" />

  <ListView android:id="@id/android:list"
    android:layout_marginLeft="5dip"
    android:layout_width="fill_parent"
    android:layout_height="wrap_content"
    android:layout_below="@id/spinner3"/>

  <TextView android:id="@id/android:empty"
    android:layout_width="wrap_content"
    android:layout_height="wrap_content"
    android:layout_marginLeft="5dip"
    android:layout_below="@id/spinner3"
    android:text="no_notes"/>

  <Button android:id="@+id/pickDate"
    android:layout_width="wrap_content"
    android:layout_height="wrap_content"
    android:layout_below="@id/spinner2"
    android:layout_toRightOf="@id/spinner3"
    android:text="change date"/>
  <Button android:id="@+id/ok"
    android:layout_width="wrap_content"
    android:layout_height="wrap_content"
    android:layout_toRightOf="@id/pickDate"
    android:layout_below="@id/spinner2"
    android:text="ok go!"/>


</RelativeLayout>
```

(a)

(b)

(c)

Fig. 6. (a) Default conditions.
(b) Default date. (c) Program list.

Code Listing 13. Creating an Adapter

```
s1 = (Spinner) findViewById(R.id.spinner1);
ArrayAdapter<CharSequence> adapter =
      ArrayAdapter.createFromResource(this, R.array.stations,
      android.R.layout.simple_spinner_item);
adapter.setDropDownViewResource(android.R.layout.simple_
      pinner_dropdown_item);
s1.setAdapter(adapter);
setListAdapter(new ArrayAdapter<String>(this,
      android.R.layout.simple_list_item_1, mStrings));
```

Code Listing 14. Obtain a pointer

```
private Button button_ok;
protected void onCreate(Bundle savedInstanceState) {
    super.onCreate(savedInstanceState);
    setContentView(R.layout.list1);
    button_ok = (Button) findViewById(R.id.ok);
    ... ...}
public class Mobile_EPG extends ListActivity {
    private Button.OnClickListener dis_list = new
      Button.OnClickListener(){public void onClick(View v){
      ... ...      }}};
protected void onCreate(Bundle savedInstanceState) {
    ... ...
    button_ok = (Button) findViewById(R.id.ok);
    button_ok.setOnClickListener(dis_list);}
```

Here a pointer to the *button_ok* element is acquired by calling the *findViewById* function. This allows us to add a click listener so that the user is notified when the button is selected on a touch screen with a stylus or by the center softkey. The *OnClickListener* class is defined inline (see Code listing 14). In this inline definition, when notified that a selection has occurred, the code in function *onclick* will be run.

4.5 EPG Database Operations

After the EPG user has selected the conditions, the EPG application will form the SQL statement. If the network status is HOME, this application will use function *udpclt.sentMsg*(condition) to send the SQL statement to the server to get feedback. If the network status is REMOTE, the EPG application will use the local database, as shown in Code Listing 15.

Code Listing 15. Database operations

```
mDbHelper = new NotesDbAdapter(this);
 mDbHelper.open();
 ... ...
mNotesCursor = mDbHelper.fetchAllNotes(sql_station, sql_date,
        sql_time_range, sql_category);
Cursor fetchAllNotes(String station, String date, String time_range,
        String category ){
    ... ...
    return mDb.query(DATABASE_TABLE, new String[] {KEY_ROWID,
        KEY_TITLE, KEY_STATION, KEY_TIME, KEY_DATE, KEY_CATEGORY},
        WHERE, null, null, null, null);}
```

4.6 The Handoff Operation

The EPG recommender handoff operation is enabled by monitoring the network modes of the mobile device. When the mobile status is HOME and the message module is waiting for the EPG query result, if a timeout occurs, the network link enters into *LINK_UNSTABLE* state to indicate that the EPG server becomes unavailable temporarily (see Code Listing 5). The function *udpclt.sentMsg* returns the string *LINK_UNSTABLE* to the *Mobile_EPG* class (see Code Listing 16). When this status is checked, the *UdpClt* class will choose the built-in EPG recommender engine. In such a manner, the handoff from HOME to REMOTE (mobile) is realized.

Code Listing 16. Timeout handling

```
try {socket.receive(packet);}
catch (SocketTimeoutException ste) {
    // When time out occurs, mobiletv can use the lite epg engine!
    // So we can just return a null or some string to indicate the
    // server is unavailable now!
    socket.close();
    return LINK_UNSTABLE;}
```

4.7 User Profile

A user profile is used for recording the user preferences associated with the recommender engine. The data structure of the user profile, *userprofiletemp*, is illustrated in Table 2. Because the user profile needs to record which categories and programs the user hits and likes, the table has the *hitTimes* and *like* fields do this task. The *hitTimes* field is responsible for recording how many times the user clicks on the EPG item displayed in the Android *listview* widget, while the *like* field is for recording whether the user likes this program when he or she browsing the EPG list (1 represents like, 0 represents dislike).

Table 2. The User Profile Table

_id	program	hitTimes	like
000	Just Shoot Me	1	1
001	Oprah Winfrey	2	0
002	Jimmy Kimmel	0	1

When an item of the program list is clicked, the function *onListItemClick* will be called. The user profile is created by this function. If there is a *userprofiletemp* table in the database, it means that the EPG application has not synchronized the user profile with the server. In this case, the program is searched which has been selected by the user in the *userprofiletemp* table. If the corresponding record is found, the *hitTimes* field will be increased by 1. If not, a new record will be inserted into this table. If the EPG application has synchronized with the server successfully, a brand new table named *userprofiletemp* will be created in the database. Next, the function *openItemDetailDialog* will start to run. It can pop up a dialog box to display the details of this program. Two buttons will be displayed in this dialog box, *like* and *dislike* respectively, as illustrated in Fig. 7. If the user clicks the like *button*, the *like* field of the corresponding record will be set to 1. Similarly, if the user clicks the *dislike* button, this field will be set to 0.

Fig. 7. Dialog box for user preference

About the timing of the user profile transfer, when the mobile device is connected to the home network again, its user profile is synchronized to the server to update its latest user profile. Here we use a variable to indicate whether to send the user profile to the server or not. If the value of *transflag* is 0, it means that there is no need to transfer the user profile to the server. Otherwise, if the value is 1, the function *synchronize* is used to send the user profile to the server, and then delete this file (by deleting the table *uesrprofiletemp* in the database) and set the value of *transflag* to 0. From the above discussion, we know that the userprofiletemp file is a table in the database, thus sending the table to the server is a difficult task. In fact we use the function *parseuserprofile* to translate the table to a XML format string and then send this string to the server before sending the recommendation request. If the recommendation list is received successfully, it means the synchronization is accomplished successfully.

5 Other Development Issues

5.1 Android Security Model

A central design point of the Android security architecture is that no application, by default, has permission to perform any operations that would adversely impact other applications, the operating system, or the user. This includes reading or writing the user's private data (such as contacts or e-mails), reading or writing another application's files, performing network access, keeping the device awake, etc.

An application's process is a secure sandbox. It cannot disrupt other applications, except by explicitly declaring the *permissions* it needs for additional capabilities not provided by the basic sandbox. The *permissions* requested can be handled by the Android platform in various ways, typically by automatically allowing or disallowing based on certificates or by prompting the user. The *permissions* required by an application are declared statically in that application, so they can be known at the installation time and will not change after that [14].

Code Listing 17. Specify the user permission

```
<manifest xmlns:android="http://schemas.android.com/apk/res/android"
          package="com.android.epg">
    <uses-permission  android:name="android.permission.INTERNET" />
</manifest>
```

A basic Android application has no *permissions* associated with it, meaning that it cannot do anything that would adversely impact the user experience or any data on the device. To make use of protected features of the device, the AndroidManifest.xml file must include one or more *<uses-permission>* tags declaring the permissions that the application needs. The EPG application that needs to open the network sockets would specify the user permission as in Code Listing 17.

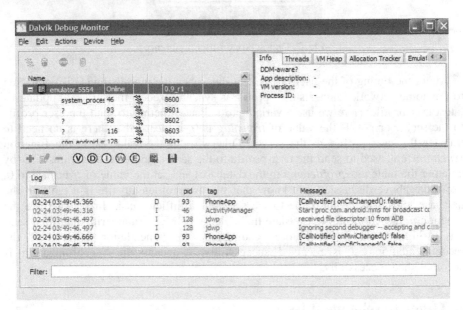

Fig. 8. The Dalvik Debug Monitor Service

5.2 Debugging Android Applications

The commonly used tools for debugging Android application are ADB and DDMS. ADB is a command-line debugging application shipped with the SDK [14]. It provides tools to browse the device, copy tools on the device, and forward ports for debugging. For example, we use DataBase Management System SharpPlus Sqlite Developer in the

EPG application to create a database named epgdb in the root directory of the developer's computer and want to send it to directory */data/data/com.android.epg /databases* on the Android's virtual machine (Dalvik). The ADB commands can be issued from a command line on the development machine. The usage is: *adb push \epgdb /data/data/com.android.epg/databases/epgdb.*

Android is shipped with a debugging tool called the Dalvik Debug Monitor Service (DDMS), which provides port-forwarding services, screen capture on the device, thread and heap information on the device, logcat, process, and radio state information, incoming call and SMS spoofing, location data spoofing, and more. DDMS is shipped in the tools/ directory of the SDK. To run it, enter this directory from a terminal/console and type ddms. DDMS will work with both the emulator and a connected device. If both are connected and running simultaneously, DDMS defaults to the emulator. Its running screen is shown in Fig. 8. Most screen images used in this chapter are captured on the emulator by selecting **Device > Screen capture...** in the menu bar, or press CTRL-S.

5.3 Signing and Publishing the Android Applications

All applications must be signed. The system will not install an application that is not signed. The Android system requires that all installed applications be digitally signed with a certificate whose private key is held by the application's developer. The system uses the certificate as a means to identify the author of an application and establishing trust relationships between applications, rather than for controlling which applications the user can install. The certificate does not need to be signed by a certificate authority: it is perfectly allowable, and typical, for Android applications to use self-signed certificates. More information about how to sign and publish an Android application is given in the Android documents [14].

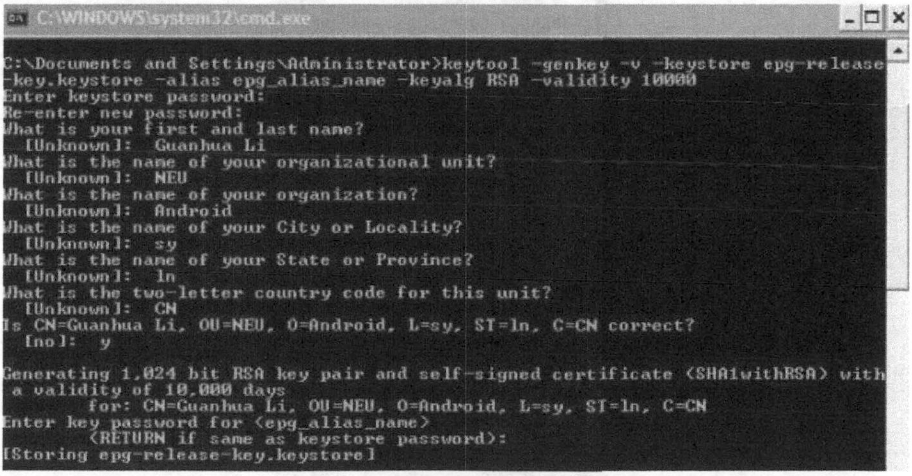

Fig. 9. Obtain a suitable private key

When the EPG application is ready for release to other users, the following steps are to be done to sign it.

1) Compile the application in release mode. Right-click the project in the Package pane of Eclipse and select **Android Tools > Export Application Package** and specify the file location for the unsigned .apk.
2) Obtain a suitable private key. An example of a Keytool command that generates a private key is shown in Fig. 9.
3) Sign the application with the private key. To sign the EPG application, run Jarsigner, referencing both the application's .apk and the keystore containing the private key with which to sign the .apk, as shown in Fig. 10.
4) Secure the private key.

As shown in Fig.11, the application mobileTV has been signed.

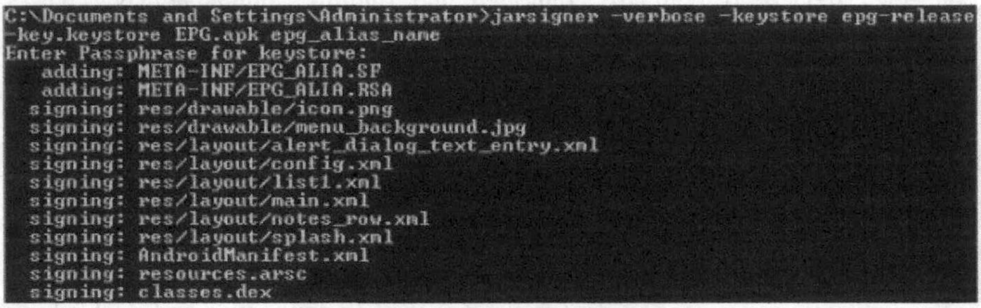

C:\Documents and Settings\Administrator>jarsigner -verbose -keystore epg-release
-key.keystore EPG.apk epg_alias_name
Enter Passphrase for keystore:
adding: META-INF/EPG_ALIA.SF
adding: META-INF/EPG_ALIA.RSA
signing: res/drawable/icon.png
signing: res/drawable/menu_background.jpg
signing: res/layout/alert_dialog_text_entry.xml
signing: res/layout/config.xml
signing: res/layout/list1.xml
signing: res/layout/main.xml
signing: res/layout/notes_row.xml
signing: res/layout/splash.xml
signing: AndroidManifest.xml
signing: resources.arsc
signing: classes.dex

Fig. 10. Sign the application with the private key

7:55 AM
Alarm Clock
Browser
Calculator
Camera
Contacts
Dev Tools
Dialer
Maps
Messaging
MobileTV
Music
Pictures
MENU

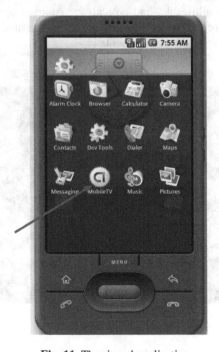

Fig. 11. The signed application

6 Concluding Remarks

A simplified EPG recommender system have been implemented on the Android platform. Key steps in developing this application are presented as an example in order to show readers how to develop a typical Android application. Our experience with the Android platform has proved the value of this new development tool for mobile devices and corresponds to those of other developers.

The development for Android is more streamlined compared to other platforms, such as Symbian [5], getting up and running with the development environment is quick and straightforward [13][20]. Compared to BREW and Java ME BREW, an activity, content handler, intent receiver, or service in Android can be written to handle nearly anything [8].

The applications developed must be tested on the actual target devices before deployment. The AAPT tool can be used to create .apk files containing the binaries and resources of Android applications. The ADB tool can be used to install the application's .apk files on a device and access the device from a command line [14]. The applications may also be deployed to an actual device using the Android Eclipse plugin.

The application software developed should be optimized to achieve the lowest power and maximum throughput before porting to actual devices, including platform level optimization and the optimization of the algorithms used in the applications [13]. This is one of our future works for the EPG example.

Acknowledgement

Authors of this book chapter wish to thank Dr. Matthew Ma at Scientific Works and Mr. Yongqiang Li at the School of Information Science and Technology of Northeastern University for their support to this work.

References

1. Adhikari, R.: HTC Adds Tattoo to Android Lineup,
    http://www.linuxinsider.com/
2. Alvarez, A.: Indepth Look at Google Android mobile phone software,
    http://techzulu.com/
3. Broersma, M.: Android invades hardware, http://www.techworld.com/
4. Cooksey, T.: Native C *GRAPHICAL* applications now working on Android emulator,
    http://groups.google.com/
5. Coulton, P., Edwards, R.C., Clemson, H.: S60 Programming: A Tutorial Guide. Wiley, Hoboken (2007)
6. Falconer, J.: 2009 Named "Year of the T-Mobile G1 Paperweight". IntoMobile (2008)
7. Freke, J.: Busybox on the G1, http://androidcommunity.com/
8. Haseman, C.: Android Essentials. Apress, Berkeley (2008)
9. Ma, M., Zhu, J., Guo, J.K.: A Recommender Framework for Electronic Programming Guide on A Mobile Device. In: Proceeding of IEEE Int. Conf. of Multimedia and Expo (2007)

10. Ma, M., Zhu, C., Tan, C., Chang, G., Zhu, J., An, Q.: A recommender handoff framework with DVB-H support on a mobile device. In: Proceeding of the First International Workshop on Mobile Multimedia (WMMP 2008), Tampa, Florida, pp. 64–71 (2008)
11. Mark, R.: Pre-Sales of T-Mobile G1 Android Hit 1.5 Million, http://www.publish.com/
12. Mazzocchi, S.: Dalvik: how Google routed around Sun's IP-based licensing restrictions on Java ME, http://www.betaversion.org/
13. Meier, R.: Professional Android Application Development. Wiley, Indianapolis (2008)
14. Open Handset Alliance, http://code.google.com/
15. Paul, R.: Developing apps for Google Android: it's a mixed bag, http://arstechnica.com/
16. Rischpater, R.: Software Development for the QUALCOMM BREW Platform. Apress, Berkeley (2003)
17. Sayer, P.: Android comes to life in Barcelona, http://www.infoworld.com/
18. Toker, A.: Skia graphics library in Chrome: First impressions, http://www.atoker.com/
19. Yacoub, H.B.: Running C++ Native Applications on Android, The Final Point, http://openhandsetmagazine.com/
20. Yacoub, H.B.: Interview with Adriano Chiaretta from iambic, http://openhandsetmagazine.com/
21. Zhu, J., Ma, M., Guo, J.K., Wang, Z.: Content Classification for Electronic Programming Guide Recommendation for a Portable Device. Int. Journal of Pattern Recognition and Artificial Intelligence 21(2), 375–395 (2007)

Author Index